Tides: A Very Short Introduction

VERY SHORT INTRODUCTIONS are for anyone wanting a stimulating and accessible way into a new subject. They are written by experts, and have been translated into more than 45 different languages.

The series began in 1995, and now covers a wide variety of topics in every discipline. The VSI library currently contains over 600 volumes—a Very Short Introduction to everything from Psychology and Philosophy of Science to American History and Relativity—and continues to grow in every subject area.

Very Short Introductions available now:

ABOLITIONISM Richard S. Newman
THE ABRAHAMIC RELIGIONS Charles L. Cohen
ACCOUNTING Christopher Nobes
ADAM SMITH Christopher J. Berry
ADOLESCENCE Peter K. Smith
ADVERTISING Winston Fletcher
AESTHETICS Bence Nanay
AFRICAN AMERICAN RELIGION Eddie S. Glaude Jr
AFRICAN HISTORY John Parker and Richard Rathbone
AFRICAN POLITICS Ian Taylor
AFRICAN RELIGIONS Jacob K. Olupona
AGEING Nancy A. Pachana
AGNOSTICISM Robin Le Poidevin
AGRICULTURE Paul Brassley and Richard Soffe
ALEXANDER THE GREAT Hugh Bowden
ALGEBRA Peter M. Higgins
AMERICAN CULTURAL HISTORY Eric Avila
AMERICAN FOREIGN RELATIONS Andrew Preston
AMERICAN HISTORY Paul S. Boyer
AMERICAN IMMIGRATION David A. Gerber
AMERICAN LEGAL HISTORY G. Edward White
AMERICAN NAVAL HISTORY Craig L. Symonds
AMERICAN POLITICAL HISTORY Donald Critchlow
AMERICAN POLITICAL PARTIES AND ELECTIONS L. Sandy Maisel
AMERICAN POLITICS Richard M. Valelly
THE AMERICAN PRESIDENCY Charles O. Jones
THE AMERICAN REVOLUTION Robert J. Allison
AMERICAN SLAVERY Heather Andrea Williams
THE AMERICAN WEST Stephen Aron
AMERICAN WOMEN'S HISTORY Susan Ware
ANAESTHESIA Aidan O'Donnell
ANALYTIC PHILOSOPHY Michael Beaney
ANARCHISM Colin Ward
ANCIENT ASSYRIA Karen Radner
ANCIENT EGYPT Ian Shaw
ANCIENT EGYPTIAN ART AND ARCHITECTURE Christina Riggs
ANCIENT GREECE Paul Cartledge
THE ANCIENT NEAR EAST Amanda H. Podany
ANCIENT PHILOSOPHY Julia Annas
ANCIENT WARFARE Harry Sidebottom
ANGELS David Albert Jones
ANGLICANISM Mark Chapman
THE ANGLO-SAXON AGE John Blair
ANIMAL BEHAVIOUR Tristram D. Wyatt
THE ANIMAL KINGDOM Peter Holland
ANIMAL RIGHTS David DeGrazia

THE ANTARCTIC Klaus Dodds
ANTHROPOCENE Erle C. Ellis
ANTISEMITISM Steven Beller
ANXIETY Daniel Freeman and Jason Freeman
THE APOCRYPHAL GOSPELS Paul Foster
APPLIED MATHEMATICS Alain Goriely
ARCHAEOLOGY Paul Bahn
ARCHITECTURE Andrew Ballantyne
ARISTOCRACY William Doyle
ARISTOTLE Jonathan Barnes
ART HISTORY Dana Arnold
ART THEORY Cynthia Freeland
ARTIFICIAL INTELLIGENCE Margaret A. Boden
ASIAN AMERICAN HISTORY Madeline Y. Hsu
ASTROBIOLOGY David C. Catling
ASTROPHYSICS James Binney
ATHEISM Julian Baggini
THE ATMOSPHERE Paul I. Palmer
AUGUSTINE Henry Chadwick
AUSTRALIA Kenneth Morgan
AUTISM Uta Frith
AUTOBIOGRAPHY Laura Marcus
THE AVANT GARDE David Cottington
THE AZTECS Davíd Carrasco
BABYLONIA Trevor Bryce
BACTERIA Sebastian G. B. Amyes
BANKING John Goddard and John O. S. Wilson
BARTHES Jonathan Culler
THE BEATS David Sterritt
BEAUTY Roger Scruton
BEHAVIOURAL ECONOMICS Michelle Baddeley
BESTSELLERS John Sutherland
THE BIBLE John Riches
BIBLICAL ARCHAEOLOGY Eric H. Cline
BIG DATA Dawn E. Holmes
BIOGRAPHY Hermione Lee
BIOMETRICS Michael Fairhurst
BLACK HOLES Katherine Blundell
BLOOD Chris Cooper
THE BLUES Elijah Wald
THE BODY Chris Shilling
THE BOOK OF COMMON PRAYER Brian Cummings
THE BOOK OF MORMON Terryl Givens
BORDERS Alexander C. Diener and Joshua Hagen
THE BRAIN Michael O'Shea
BRANDING Robert Jones
THE BRICS Andrew F. Cooper
THE BRITISH CONSTITUTION Martin Loughlin
THE BRITISH EMPIRE Ashley Jackson
BRITISH POLITICS Anthony Wright
BUDDHA Michael Carrithers
BUDDHISM Damien Keown
BUDDHIST ETHICS Damien Keown
BYZANTIUM Peter Sarris
C. S. LEWIS James Como
CALVINISM Jon Balserak
CANCER Nicholas James
CAPITALISM James Fulcher
CATHOLICISM Gerald O'Collins
CAUSATION Stephen Mumford and Rani Lill Anjum
THE CELL Terence Allen and Graham Cowling
THE CELTS BarryCunliffe
CHAOS Leonard Smith
CHARLES DICKENS Jenny Hartley
CHEMISTRY Peter Atkins
CHILD PSYCHOLOGY Usha Goswami
CHILDREN'S LITERATURE Kimberley Reynolds
CHINESE LITERATURE Sabina Knight
CHOICE THEORY Michael Allingham
CHRISTIAN ART Beth Williamson
CHRISTIAN ETHICS D. Stephen Long
CHRISTIANITY Linda Woodhead
CIRCADIAN RHYTHMS Russell Foster and Leon Kreitzman
CITIZENSHIP Richard Bellamy
CIVIL ENGINEERING David Muir Wood
CLASSICAL LITERATURE William Allan
CLASSICAL MYTHOLOGY Helen Morales
CLASSICS Mary Beard and John Henderson
CLAUSEWITZ MIchael Howard
CLIMATE Mark Maslin
CLIMATE CHANGE Mark Maslin
CLINICAL PSYCHOLOGY Susan Llewelyn and Katie Aafjes-van Doorn

COGNITIVE NEUROSCIENCE Richard Passingham
THE COLD WAR Robert McMahon
COLONIAL AMERICA Alan Taylor
COLONIAL LATIN AMERICAN LITERATURE Rolena Adorno
COMBINATORICS Robin Wilson
COMEDY Matthew Bevis
COMMUNISM Leslie Holmes
COMPARATIVE LITERATURE Ben Hutchinson
COMPLEXITY John H. Holland
THE COMPUTER Darrel Ince
COMPUTER SCIENCE Subrata Dasgupta
CONCENTRATION CAMPS Dan Stone
CONFUCIANISM Daniel K. Gardner
THE CONQUISTADORS Matthew Restall and Felipe Fernández-Armesto
CONSCIENCE Paul Strohm
CONSCIOUSNESS Susan Blackmore
CONTEMPORARY ART Julian Stallabrass
CONTEMPORARY FICTION Robert Eaglestone
CONTINENTAL PHILOSOPHY Simon Critchley
COPERNICUS Owen Gingerich
CORAL REEFS Charles Sheppard
CORPORATE SOCIAL RESPONSIBILITY Jeremy Moon
CORRUPTION Leslie Holmes
COSMOLOGY Peter Coles
COUNTRY MUSIC Richard Carlin
CRIME FICTION Richard Bradford
CRIMINAL JUSTICE Julian V. Roberts
CRIMINOLOGY Tim Newburn
CRITICAL THEORY Stephen Eric Bronner
THE CRUSADES Christopher Tyerman
CRYPTOGRAPHY Fred Piper and Sean Murphy
CRYSTALLOGRAPHY A. M. Glazer
THE CULTURAL REVOLUTION Richard Curt Kraus
DADA AND SURREALISM David Hopkins
DANTE Peter Hainsworth and David Robey
DARWIN Jonathan Howard
THE DEAD SEA SCROLLS Timothy H. Lim
DECADENCE David Weir
DECOLONIZATION Dane Kennedy
DEMOCRACY Bernard Crick
DEMOGRAPHY Sarah Harper
DEPRESSION Jan Scott and Mary Jane Tacchi
DERRIDA Simon Glendinning
DESCARTES Tom Sorell
DESERTS Nick Middleton
DESIGN John Heskett
DEVELOPMENT Ian Goldin
DEVELOPMENTAL BIOLOGY Lewis Wolpert
THE DEVIL Darren Oldridge
DIASPORA Kevin Kenny
DICTIONARIES Lynda Mugglestone
DINOSAURS David Norman
DIPLOMACY Joseph M. Siracusa
DOCUMENTARY FILM Patricia Aufderheide
DREAMING J. Allan Hobson
DRUGS Les Iversen
DRUIDS Barry Cunliffe
DYNASTY Jeroen Duindam
DYSLEXIA Margaret J. Snowling
EARLY MUSIC Thomas Forrest Kelly
THE EARTH Martin Redfern
EARTH SYSTEM SCIENCE Tim Lenton
ECONOMICS Partha Dasgupta
EDUCATION Gary Thomas
EGYPTIAN MYTH Geraldine Pinch
EIGHTEENTH-CENTURY BRITAIN Paul Langford
THE ELEMENTS Philip Ball
EMOTION Dylan Evans
EMPIRE Stephen Howe
ENERGY SYSTEMS Nick Jenkins
ENGELS Terrell Carver
ENGINEERING David Blockley
THE ENGLISH LANGUAGE Simon Horobin
ENGLISH LITERATURE Jonathan Bate
THE ENLIGHTENMENT John Robertson
ENTREPRENEURSHIP Paul Westhead and Mike Wright
ENVIRONMENTAL ECONOMICS Stephen Smith

ENVIRONMENTAL ETHICS Robin Attfield
ENVIRONMENTAL LAW Elizabeth Fisher
ENVIRONMENTAL POLITICS Andrew Dobson
EPICUREANISM Catherine Wilson
EPIDEMIOLOGY Rodolfo Saracci
ETHICS Simon Blackburn
ETHNOMUSICOLOGY Timothy Rice
THE ETRUSCANS Christopher Smith
EUGENICS Philippa Levine
THE EUROPEAN UNION Simon Usherwood and John Pinder
EUROPEAN UNION LAW Anthony Arnull
EVOLUTION Brian and Deborah Charlesworth
EXISTENTIALISM Thomas Flynn
EXPLORATION Stewart A. Weaver
EXTINCTION Paul B. Wignall
THE EYE Michael Land
FAIRY TALE Marina Warner
FAMILY LAW Jonathan Herring
FASCISM Kevin Passmore
FASHION Rebecca Arnold
FEDERALISM Mark J. Rozell and Clyde Wilcox
FEMINISM Margaret Walters
FILM Michael Wood
FILM MUSIC Kathryn Kalinak
FILM NOIR James Naremore
THE FIRST WORLD WAR Michael Howard
FOLK MUSIC Mark Slobin
FOOD John Krebs
FORENSIC PSYCHOLOGY David Canter
FORENSIC SCIENCE Jim Fraser
FORESTS Jaboury Ghazoul
FOSSILS Keith Thomson
FOUCAULT Gary Gutting
THE FOUNDING FATHERS R. B. Bernstein
FRACTALS Kenneth Falconer
FREE SPEECH Nigel Warburton
FREE WILL Thomas Pink
FREEMASONRY Andreas Önnerfors
FRENCH LITERATURE John D. Lyons
THE FRENCH REVOLUTION William Doyle
FREUD Anthony Storr
FUNDAMENTALISM Malise Ruthven
FUNGI Nicholas P. Money
THE FUTURE Jennifer M. Gidley
GALAXIES John Gribbin
GALILEO Stillman Drake
GAME THEORY Ken Binmore
GANDHI Bhikhu Parekh
GARDEN HISTORY Gordon Campbell
GENES Jonathan Slack
GENIUS Andrew Robinson
GENOMICS John Archibald
GEOFFREY CHAUCER David Wallace
GEOGRAPHY John Matthews and David Herbert
GEOLOGY Jan Zalasiewicz
GEOPHYSICS William Lowrie
GEOPOLITICS Klaus Dodds
GERMAN LITERATURE Nicholas Boyle
GERMAN PHILOSOPHY Andrew Bowie
GLACIATION David J. A. Evans
GLOBAL CATASTROPHES Bill McGuire
GLOBAL ECONOMIC HISTORY Robert C. Allen
GLOBALIZATION Manfred Steger
GOD John Bowker
GOETHE Ritchie Robertson
THE GOTHIC Nick Groom
GOVERNANCE Mark Bevir
GRAVITY Timothy Clifton
THE GREAT DEPRESSION AND THE NEW DEAL Eric Rauchway
HABERMAS James Gordon Finlayson
THE HABSBURG EMPIRE Martyn Rady
HAPPINESS Daniel M. Haybron
THE HARLEM RENAISSANCE Cheryl A. Wall
THE HEBREW BIBLE AS LITERATURE Tod Linafelt
HEGEL Peter Singer
HEIDEGGER Michael Inwood
THE HELLENISTIC AGE Peter Thonemann
HEREDITY John Waller
HERMENEUTICS Jens Zimmermann
HERODOTUS Jennifer T. Roberts
HIEROGLYPHS Penelope Wilson
HINDUISM Kim Knott
HISTORY John H. Arnold

THE HISTORY OF ASTRONOMY Michael Hoskin
THE HISTORY OF CHEMISTRY William H. Brock
THE HISTORY OF CHILDHOOD James Marten
THE HISTORY OF CINEMA Geoffrey Nowell-Smith
THE HISTORY OF LIFE Michael Benton
THE HISTORY OF MATHEMATICS Jacqueline Stedall
THE HISTORY OF MEDICINE William Bynum
THE HISTORY OF PHYSICS J. L. Heilbron
THE HISTORY OF TIME Leofranc Holford-Strevens
HIV AND AIDS Alan Whiteside
HOBBES Richard Tuck
HOLLYWOOD Peter Decherney
THE HOLY ROMAN EMPIRE Joachim Whaley
HOME Michael Allen Fox
HOMER Barbara Graziosi
HORMONES Martin Luck
HUMAN ANATOMY Leslie Klenerman
HUMAN EVOLUTION Bernard Wood
HUMAN RIGHTS Andrew Clapham
HUMANISM Stephen Law
HUME A. J. Ayer
HUMOUR Noël Carroll
THE ICE AGE Jamie Woodward
IDENTITY Florian Coulmas
IDEOLOGY Michael Freeden
THE IMMUNE SYSTEM Paul Klenerman
INDIAN CINEMA Ashish Rajadhyaksha
INDIAN PHILOSOPHY Sue Hamilton
THE INDUSTRIAL REVOLUTION Robert C. Allen
INFECTIOUS DISEASE Marta L. Wayne and Benjamin M. Bolker
INFINITY Ian Stewart
INFORMATION Luciano Floridi
INNOVATION Mark Dodgson and David Gann
INTELLECTUAL PROPERTY Siva Vaidhyanathan
INTELLIGENCE Ian J. Deary
INTERNATIONAL LAW Vaughan Lowe
INTERNATIONAL MIGRATION Khalid Koser
INTERNATIONAL RELATIONS Paul Wilkinson
INTERNATIONAL SECURITY Christopher S. Browning
IRAN Ali M. Ansari
ISLAM Malise Ruthven
ISLAMIC HISTORY Adam Silverstein
ISOTOPES Rob Ellam
ITALIAN LITERATURE Peter Hainsworth and David Robey
JESUS Richard Bauckham
JEWISH HISTORY David N. Myers
JOURNALISM Ian Hargreaves
JUDAISM Norman Solomon
JUNG Anthony Stevens
KABBALAH Joseph Dan
KAFKA Ritchie Robertson
KANT Roger Scruton
KEYNES Robert Skidelsky
KIERKEGAARD Patrick Gardiner
KNOWLEDGE Jennifer Nagel
THE KORAN Michael Cook
LAKES Warwick F. Vincent
LANDSCAPE ARCHITECTURE Ian H. Thompson
LANDSCAPES AND GEOMORPHOLOGY Andrew Goudie and Heather Viles
LANGUAGES Stephen R. Anderson
LATE ANTIQUITY Gillian Clark
LAW Raymond Wacks
THE LAWS OF THERMODYNAMICS Peter Atkins
LEADERSHIP Keith Grint
LEARNING Mark Haselgrove
LEIBNIZ Maria Rosa Antognazza
LEO TOLSTOY Liza Knapp
LIBERALISM Michael Freeden
LIGHT Ian Walmsley
LINCOLN Allen C. Guelzo
LINGUISTICS Peter Matthews
LITERARY THEORY Jonathan Culler
LOCKE John Dunn
LOGIC Graham Priest
LOVE Ronald de Sousa

MACHIAVELLI Quentin Skinner
MADNESS Andrew Scull
MAGIC Owen Davies
MAGNA CARTA Nicholas Vincent
MAGNETISM Stephen Blundell
MALTHUS Donald Winch
MAMMALS T. S. Kemp
MANAGEMENT John Hendry
MAO Delia Davin
MARINE BIOLOGY Philip V. Mladenov
THE MARQUIS DE SADE John Phillips
MARTIN LUTHER Scott H. Hendrix
MARTYRDOM Jolyon Mitchell
MARX Peter Singer
MATERIALS Christopher Hall
MATHEMATICAL FINANCE Mark H. A. Davis
MATHEMATICS Timothy Gowers
MATTER Geoff Cottrell
THE MEANING OF LIFE Terry Eagleton
MEASUREMENT David Hand
MEDICAL ETHICS Michael Dunn and Tony Hope
MEDICAL LAW Charles Foster
MEDIEVAL BRITAIN John Gillingham and Ralph A. Griffiths
MEDIEVAL LITERATURE Elaine Treharne
MEDIEVAL PHILOSOPHY John Marenbon
MEMORY Jonathan K. Foster
METAPHYSICS Stephen Mumford
METHODISM William J. Abraham
THE MEXICAN REVOLUTION Alan Knight
MICHAEL FARADAY Frank A. J. L. James
MICROBIOLOGY Nicholas P. Money
MICROECONOMICS Avinash Dixit
MICROSCOPY Terence Allen
THE MIDDLE AGES Miri Rubin
MILITARY JUSTICE Eugene R. Fidell
MILITARY STRATEGY Antulio J. Echevarria II
MINERALS David Vaughan
MIRACLES Yujin Nagasawa
MODERN ARCHITECTURE Adam Sharr
MODERN ART David Cottington
MODERN CHINA Rana Mitter
MODERN DRAMA Kirsten E. Shepherd-Barr
MODERN FRANCE Vanessa R. Schwartz
MODERN INDIA Craig Jeffrey
MODERN IRELAND Senia Pašeta
MODERN ITALY Anna Cento Bull
MODERN JAPAN Christopher Goto-Jones
MODERN LATIN AMERICAN LITERATURE Roberto González Echevarría
MODERN WAR Richard English
MODERNISM Christopher Butler
MOLECULAR BIOLOGY Aysha Divan and Janice A. Royds
MOLECULES Philip Ball
MONASTICISM Stephen J. Davis
THE MONGOLS Morris Rossabi
MOONS David A. Rothery
MORMONISM Richard Lyman Bushman
MOUNTAINS Martin F. Price
MUHAMMAD Jonathan A. C. Brown
MULTICULTURALISM Ali Rattansi
MULTILINGUALISM John C. Maher
MUSIC Nicholas Cook
MYTH Robert A. Segal
NAPOLEON David Bell
THE NAPOLEONIC WARS Mike Rapport
NATIONALISM Steven Grosby
NATIVE AMERICAN LITERATURE Sean Teuton
NAVIGATION Jim Bennett
NAZI GERMANY Jane Caplan
NELSON MANDELA Elleke Boehmer
NEOLIBERALISM Manfred Steger and Ravi Roy
NETWORKS Guido Caldarelli and Michele Catanzaro
THE NEW TESTAMENT Luke Timothy Johnson
THE NEW TESTAMENT AS LITERATURE Kyle Keefer
NEWTON Robert Iliffe
NIETZSCHE Michael Tanner
NINETEENTH-CENTURY BRITAIN Christopher Harvie and H. C. G. Matthew

THE NORMAN CONQUEST George Garnett
NORTH AMERICAN INDIANS Theda Perdue and Michael D. Green
NORTHERN IRELAND Marc Mulholland
NOTHING Frank Close
NUCLEAR PHYSICS Frank Close
NUCLEAR POWER Maxwell Irvine
NUCLEAR WEAPONS Joseph M. Siracusa
NUMBERS Peter M. Higgins
NUTRITION David A. Bender
OBJECTIVITY Stephen Gaukroger
OCEANS Dorrik Stow
THE OLD TESTAMENT Michael D. Coogan
THE ORCHESTRA D. Kern Holoman
ORGANIC CHEMISTRY Graham Patrick
ORGANIZATIONS Mary Jo Hatch
ORGANIZED CRIME Georgios A. Antonopoulos and Georgios Papanicolaou
ORTHODOX CHRISTIANITY A. Edward Siecienski
PAGANISM Owen Davies
PAIN Rob Boddice
THE PALESTINIAN-ISRAELI CONFLICT Martin Bunton
PANDEMICS Christian W. McMillen
PARTICLE PHYSICS Frank Close
PAUL E. P. Sanders
PEACE Oliver P. Richmond
PENTECOSTALISM William K. Kay
PERCEPTION Brian Rogers
THE PERIODIC TABLE Eric R. Scerri
PHILOSOPHY Edward Craig
PHILOSOPHY IN THE ISLAMIC WORLD Peter Adamson
PHILOSOPHY OF BIOLOGY Samir Okasha
PHILOSOPHY OF LAW Raymond Wacks
PHILOSOPHY OF RELIGION Tim Bayne
PHILOSOPHY OF SCIENCE Samir Okasha
PHOTOGRAPHY Steve Edwards
PHYSICAL CHEMISTRY Peter Atkins
PHYSICS Sidney Perkowitz
PILGRIMAGE Ian Reader
PLAGUE Paul Slack
PLANETS David A. Rothery
PLANTS Timothy Walker
PLATE TECTONICS Peter Molnar
PLATO Julia Annas
POETRY Bernard O'Donoghue
POLITICAL PHILOSOPHY David Miller
POLITICS Kenneth Minogue
POPULISM Cas Mudde and Cristóbal Rovira Kaltwasser
POSTCOLONIALISM Robert Young
POSTMODERNISM Christopher Butler
POSTSTRUCTURALISM Catherine Belsey
POVERTY Philip N. Jefferson
PREHISTORY Chris Gosden
PRESOCRATIC PHILOSOPHY Catherine Osborne
PRIVACY Raymond Wacks
PROBABILITY John Haigh
PROGRESSIVISM Walter Nugent
PROJECTS Andrew Davies
PROTESTANTISM Mark A. Noll
PSYCHIATRY Tom Burns
PSYCHOANALYSIS Daniel Pick
PSYCHOLOGY Gillian Butler and Freda McManus
PSYCHOLOGY OF MUSIC Elizabeth Hellmuth Margulis
PSYCHOPATHY Essi Viding
PSYCHOTHERAPY Tom Burns and Eva Burns-Lundgren
PUBLIC ADMINISTRATION Stella Z. Theodoulou and Ravi K. Roy
PUBLIC HEALTH Virginia Berridge
PURITANISM Francis J. Bremer
THE QUAKERS Pink Dandelion
QUANTUM THEORY John Polkinghorne
RACISM Ali Rattansi
RADIOACTIVITY Claudio Tuniz
RASTAFARI Ennis B. Edmonds
READING Belinda Jack
THE REAGAN REVOLUTION Gil Troy
REALITY Jan Westerhoff
THE REFORMATION Peter Marshall
RELATIVITY Russell Stannard

RELIGION IN AMERICA Timothy Beal
THE RENAISSANCE Jerry Brotton
RENAISSANCE ART
Geraldine A. Johnson
REPTILES T. S. Kemp
REVOLUTIONS Jack A. Goldstone
RHETORIC Richard Toye
RISK Baruch Fischhoff and John Kadvany
RITUAL Barry Stephenson
RIVERS Nick Middleton
ROBOTICS Alan Winfield
ROCKS Jan Zalasiewicz
ROMAN BRITAIN Peter Salway
THE ROMAN EMPIRE
Christopher Kelly
THE ROMAN REPUBLIC
David M. Gwynn
ROMANTICISM Michael Ferber
ROUSSEAU Robert Wokler
RUSSELL A. C. Grayling
RUSSIAN HISTORY Geoffrey Hosking
RUSSIAN LITERATURE Catriona Kelly
THE RUSSIAN REVOLUTION
S. A. Smith
SAINTS Simon Yarrow
SAVANNAS Peter A. Furley
SCEPTICISM Duncan Pritchard
SCHIZOPHRENIA Chris Frith and
Eve Johnstone
SCHOPENHAUER
Christopher Janaway
SCIENCE AND RELIGION
Thomas Dixon
SCIENCE FICTION David Seed
THE SCIENTIFIC REVOLUTION
Lawrence M. Principe
SCOTLAND Rab Houston
SECULARISM Andrew Copson
SEXUAL SELECTION Marlene Zuk
and Leigh W. Simmons
SEXUALITY Véronique Mottier
SHAKESPEARE'S COMEDIES
Bart van Es
SHAKESPEARE'S SONNETS
AND POEMS Jonathan F. S. Post
SHAKESPEARE'S TRAGEDIES
Stanley Wells
SIKHISM Eleanor Nesbitt
THE SILK ROAD James A. Millward
SLANG Jonathon Green
SLEEP Steven W. Lockley and
Russell G. Foster
SOCIAL AND CULTURAL
ANTHROPOLOGY
John Monaghan and Peter Just
SOCIAL PSYCHOLOGY Richard J. Crisp
SOCIAL WORK Sally Holland and
Jonathan Scourfield
SOCIALISM Michael Newman
SOCIOLINGUISTICS John Edwards
SOCIOLOGY Steve Bruce
SOCRATES C. C. W. Taylor
SOUND Mike Goldsmith
SOUTHEAST ASIA James R. Rush
THE SOVIET UNION Stephen Lovell
THE SPANISH CIVIL WAR
Helen Graham
SPANISH LITERATURE Jo Labanyi
SPINOZA Roger Scruton
SPIRITUALITY Philip Sheldrake
SPORT Mike Cronin
STARS Andrew King
STATISTICS David J. Hand
STEM CELLS Jonathan Slack
STOICISM Brad Inwood
STRUCTURAL ENGINEERING
David Blockley
STUART BRITAIN John Morrill
SUPERCONDUCTIVITY
Stephen Blundell
SYMMETRY Ian Stewart
SYNAESTHESIA Julia Simner
SYNTHETIC BIOLOGY Jamie A. Davies
TAXATION Stephen Smith
TEETH Peter S. Ungar
TELESCOPES Geoff Cottrell
TERRORISM Charles Townshend
THEATRE Marvin Carlson
THEOLOGY David F. Ford
THINKING AND REASONING
Jonathan St B. T. Evans
THOMAS AQUINAS Fergus Kerr
THOUGHT Tim Bayne
TIBETAN BUDDHISM
Matthew T. Kapstein
TIDES David George Bowers and
Emyr Martyn Roberts
TOCQUEVILLE Harvey C. Mansfield
TRAGEDY Adrian Poole
TRANSLATION Matthew Reynolds

THE TREATY OF VERSAILLES Michael S. Neiberg
THE TROJAN WAR Eric H. Cline
TRUST Katherine Hawley
THE TUDORS John Guy
TWENTIETH-CENTURY BRITAIN Kenneth O. Morgan
TYPOGRAPHY Paul Luna
THE UNITED NATIONS Jussi M. Hanhimäki
UNIVERSITIES AND COLLEGES David Palfreyman and Paul Temple
THE U.S. CONGRESS Donald A. Ritchie
THE U.S. CONSTITUTION David J. Bodenhamer
THE U.S. SUPREME COURT Linda Greenhouse
UTILITARIANISM Katarzyna de Lazari-Radek and Peter Singer
UTOPIANISM Lyman Tower Sargent
VETERINARY SCIENCE James Yeates
THE VIKINGS Julian D. Richards
VIRUSES Dorothy H. Crawford
VOLTAIRE Nicholas Cronk
WAR AND TECHNOLOGY Alex Roland
WATER John Finney
WAVES Mike Goldsmith
WEATHER Storm Dunlop
THE WELFARE STATE David Garland
WILLIAM SHAKESPEARE Stanley Wells
WITCHCRAFT Malcolm Gaskill
WITTGENSTEIN A. C. Grayling
WORK Stephen Fineman
WORLD MUSIC Philip Bohlman
THE WORLD TRADE ORGANIZATION Amrita Narlikar
WORLD WAR II Gerhard L. Weinberg
WRITING AND SCRIPT Andrew Robinson
ZIONISM Michael Stanislawski

Available soon:

TOPOLOGY Richard Earl
KOREA Michael J. Seth
TRIGONOMETRY Glen Van Brummelen
BIOGEOGRAPHY Mark V. Lomolino
RENEWABLE ENERGY Nick Jelley

For more information visit our website

www.oup.com/vsi/

David George Bowers and
Emyr Martyn Roberts

TIDES

A Very Short Introduction

OXFORD
UNIVERSITY PRESS

Great Clarendon Street, Oxford, OX2 6DP,
United Kingdom

Oxford University Press is a department of the University of Oxford.
It furthers the University's objective of excellence in research, scholarship,
and education by publishing worldwide. Oxford is a registered trade mark of
Oxford University Press in the UK and in certain other countries

First edition published in 2019

Published in the United States of America by Oxford University Press
198 Madison Avenue, New York, NY 10016, United States of America

British Library Cataloguing in Publication Data
Data available

Library of Congress Control Number: 2019946776

ISBN 978–0–19–882663–7

Printed and bound by
CPI Group (UK) Ltd, Croydon, CR0 4YY

For Dylan
and for all who are curious about the tide

Contents

Preface

Since time immemorial, people have stood on the shore and wondered about the tide. The cause of the twice-daily rise and fall of sea level has attracted the attention of some of the world's greatest scientists. To begin with, enquiry was fundamental (why are there two tides a day?), and practical (how can we predict it?). Today we realize that the tide is essential for making our planet the way it is. It is part of the process that controls the global climate and it helps make the oceans teem with life. At a local level, the tide can also be *spectacular*: there is nothing quite like seeing a tidal bore travelling up a river accompanied by a train of surfers.

This book is written by two scientists who continue to be surprised and intrigued by what the tide can do. The book explains how the tide is made, and how it can be measured and predicted. There are sections on tides in coastal waters, tidal bores and related phenomena, tidal friction and the slowing of the Earth's spin, and on the tide as a great mixer of our seas and oceans. The book finishes by considering what is next in tidal studies, in our own oceans and elsewhere in the solar system. These are exciting times; space probes are due to study the possibility of life in the tide-warmed oceans on the moons of Jupiter and Saturn.

Tides: A Very Short Introduction is intended for those readers who are curious about how the tide works and why it matters to our planet. It is aimed at students and non-specialized readers wanting a succinct guide to the subject. The book avoids mathematics, instead using physical arguments, analogies, and illustrations to make a point. Examples of tides and tide-related processes are given from around the world.

Acknowledgements

This book would not have been possible without the generous help of many people. We are extremely grateful to Latha Menon of the Oxford University Press for championing the idea in the beginning and to Ernest Naylor for his guidance. Further information and advice was provided by Anne-Christin Schulz of the University of Oldenburg, Germany, and Christopher Jones and Joanne Crewdson of the UK Hydrographic Office. Jenny Nugée, publication editor for VSIs, patiently guided us through the business of producing this book. Eric Jones, who has seen most of the tidal bores on our planet, offered his insights on Chapter 5. Phil Woodworth read a draft version of the book and made invaluable suggestions for improvement. The book is much improved as a result of acting on those. Bangor University, Universitetet i Bergen, and the European Union are thanked for their support.

Data for some of the illustrations were taken from the NOAA tides and currents website. Satellite data used in Chapter 7 were received and processed by the UK's NERC Earth Observation Data Acquisition and Analysis Service (NEODAAS) at Dundee University and Plymouth Marine Laboratory (http://www.neodaas.ac.uk). SeaWiFS data were provided courtesy of the NASA SeaWiFS project and Orbital Sciences Corporation. We are grateful to the Royal Society for the Protection of Birds for the data used to draw Figure 13.

David Bowers has, over the course of many years, enjoyed discussions about tides and tide-related phenomena memorably with Katherine Braithwaite, John Bye, John Brubaker, Carl Friedrichs, Mattias Green, Antonio Hoguane, Geof Lennon, Rick Nunes, Tom Rippeth, Larry Sanford, John Simpson, Martin White, and Phil Woodworth, amongst others. The fruits of those discussions adorn these pages. Faith Bowers kindly allowed her husband to disappear into his study for many hours to write his part of this book and still had the patience to tell him gracefully where he could make improvements.

Martyn Roberts is grateful to Malen for her careful reading of his contribution to this book, and to Dylan (a name derived from the Welsh word for tide) for sharing his daddy with both a computer and the sea. Kate Johnson, Howard Jukes, and Iris Verhagen are thanked for chasing tidal bores. Victoria Johnson's literary knowledge was very helpful and is surely unsurpassed. Hans Tore Rapp and Furu Mienis are thanked for all things deep sea. This book was in part inspired by conversations about the tides over many years: the Brothers Grant (Meilyr and Aeddan), Mervyn, Susan, and Geraint Roberts, and David Broadbent are responsible, although probably none can remember exactly how, where, and when.

List of illustrations

Chapter 1
Watching the tide

The tide is the ocean's response to the gravitational pull of the Moon and the Sun. More exactly, as we shall see, it is the *variation* in this pull over the surface of the Earth that makes tides. There are also tides in the atmosphere and in the solid earth, but it is only in the ocean that the tide has practical importance. On some coasts, tides are insignificant but at others the level of the sea rises and falls twice a day by several metres. At these places the tide is important for fishers bringing their boats home from sea, dog-walkers on the beach, search-and-rescue teams, anglers digging for bait-worms, enormous super-tankers bringing oil to port, archaeologists looking for submerged wrecks, and indeed everyone who lives and works by the sea.

As well as the vertical movement of the tide, there is an associated back-and-forth, mostly gentle, motion of the waters that fill the ocean. When the tidal flow is squeezed through a narrow sea strait, however, the currents can become far from gentle and create dramatic features such as *hydraulic jumps* and whirlpools. Every drop of seawater is involved in tidal motion: it is the greatest synchronized movement of matter on our planet.

The tide at the coast

Most people first come across the tide at the coast. The tide drives the edge of the sea in and out over a beach or up and down a cliff face; the shape and appearance of the shore change with the tide. Beach users have to be careful that they are not cut off by the rising sea. The advance of the tide up a beach is usually quite slow, slower than walking pace, but the danger lies in the sea filling channels between you and dry land. Moreover, there are flat coastlines where the tide does come in fast and catches people out, sometimes with fatal consequences. Tides create special features such as tidal islands which can be walked to at low water but which are surrounded by the sea when the tide comes in. In Robert Louis Stevenson's story *Kidnapped*, the hero thinks he is trapped on an island until he realizes he can walk to the mainland at low tide. Less well-known are tidal lakes, depressions in the beach which fill and empty as the tide percolates underground through the beach material.

An impressive example of a coastline which changes dramatically with the state of the tide is the southern part of the North Sea in Europe—the Wadden Sea—where a coastal strip running several miles offshore is alternately covered and uncovered by the sea. At low water there is a wide expanse of sandbanks stretching as far as the eye can see. The banks are cut here and there by channels, some of which are navigable and are marked by poles. As the tide rises, the sea flows first up these channels and then over the tops of the banks. In Erskine Childers' classic sea story, *The Riddle of the Sands*, the advance of the tide in the Wadden Sea is described:

> I waited on deck and watched the death-throes of the suffocating sands under the relentless onset of the sea. The last strongholds were battered, stormed and overwhelmed; the tumult of sounds sank as the sea swept victoriously over the whole expanse.

The importance of the tide

For centuries, knowledge of the tide has been critical for safe navigation of ships in coastal waters. The Ancient Greek and Roman civilizations weren't too concerned with the tide—with a few exceptions, Mediterranean tides are small. However, when the great age of European exploration began and northern ports of Europe grew in importance, knowledge of the tides came to be of commercial and military significance. Ships large enough to sail the Atlantic were now using harbours that became very shallow at low tide. A foul wind may hold back a ship entering Rotterdam or Liverpool, but getting the tide wrong could leave it stranded on a sandbank or breaking its back on a rock.

The need to know how much water is under a ship's keel is even more pressing today with the trend for super-sized cargo ships, tankers, and cruise liners. The launch of a large ship into a shallow estuary is a particularly tricky business. In 2017, the UK launched the aircraft carrier, *Queen Elizabeth* into the River Forth. A high tide was needed for the launch so that there would be enough water to float the ship but then it had to wait for a low tide to safely navigate under the three bridges that lay in its path to the sea.

The tide is a great powerhouse of our planet: only sunlight and the nuclear heat of the Earth's core deliver more energy. The tide has mechanical energy: potential energy which is stored as the tide rises and kinetic energy in tidal currents. In the Bay of Fundy, Canada, the sea surface rises and falls, twice a day, by over 12 metres. The energy required to lift all the water in the Bay from low to high tide is about 10^{16} joules and this energy is released when the tide falls again. If all the released energy could be captured, the power output would be equivalent to a hundred medium-sized nuclear power stations. Unlike the Sun (and the wind), the tide is a reliable source of renewable energy which gives as freely in winter as it does in summer.

In recent years we have come to realize that tides are important in subtle ways. Tides *mix* the sea. They stir the Sun's heat down into the waters around the coast, creating a store of summer heat which is released in winter to smooth out the seasons. Below the surface of the deep ocean, tidal energy also mixes the Sun's heat downwards, gently warming abyssal waters. This process is essential for maintaining the vertical circulation of the ocean and pumping heat from the tropics to the poles.

Tidal rhythms

The times and levels of low and high tide vary from place to place, but the tide everywhere has the same set of rhythms, produced by the motion of the Earth, Moon, and Sun. Tide tables are made by seeking out these rhythms, or *harmonics*, and projecting them into the future; we describe how this is done in Chapter 3. It is, however, quite easy to make rough and ready predictions—say the time of high tide next Wednesday—once the basics of the tidal rhythms have been grasped.

The most common rhythm is the *semi-diurnal tide* in which high waters occur at intervals of twelve hours and twenty-five minutes on average. Semi-diurnal tides are found throughout Europe, the east coast of North America and most of the African coastline. As an example, Figure 1 shows a plot of water level against time (a tidal curve) on two consecutive days at Juneau, a city on the eastern shore of the north Pacific.

On both days in this figure there are two high tides and two low tides, occurring later on the second day than the first. The time of a high tide advances from one day to the next by, on average, fifty minutes. This is the golden rule of the rough-and-ready school of tidal prediction. If a high tide today is at 10 am, then expect it tomorrow at about 10.50 am, the day after that at 11.40 am, and so on.

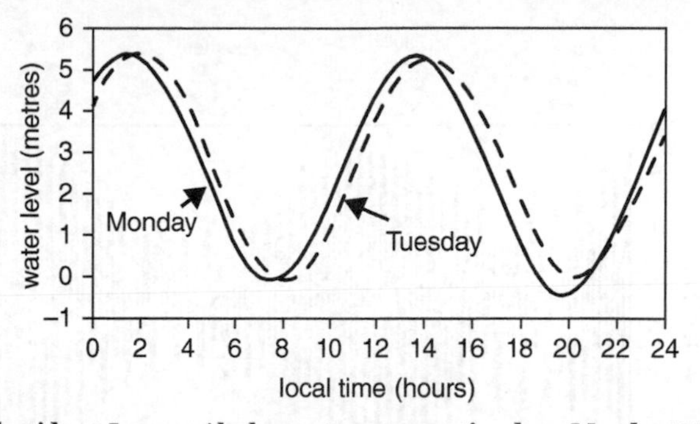

1. The tide at Juneau, Alaska, on two consecutive days, Monday 13th and Tuesday 14th of March 2017.

This daily advance in the time of high tide is a result of the orbital motion of the Moon about the spinning Earth. In twenty-four hours, the Earth turns such that the Sun returns to the same place in the sky. The Moon, however, moves some way around its orbit during this time, and the Earth has to turn a little more to catch up with it. The average interval between times when the Moon is in the same place in the sky is twenty-four hours and fifty minutes—a period called a lunar day. The time of high tide advances by fifty minutes each day to match this movement of the Moon. The fact that there are *two* high tides in a lunar day takes some explaining and we return to this in Chapter 2.

After fifteen days the time of high water has advanced by about 15x50 minutes or twelve and a half hours. If we start on a day when high tide is at 10 am, then fifteen days later, high tide will be around 10.30 pm and the *previous* high tide will now be at about 10 am. The time of high water therefore goes through a complete cycle, returning more or less back to where it started, every fifteen days. The exact timescale for this repeating cycle is in fact a few hours less than a full fifteen days. It is the time taken for the phase of the Moon to change from new to full, or vice-versa.

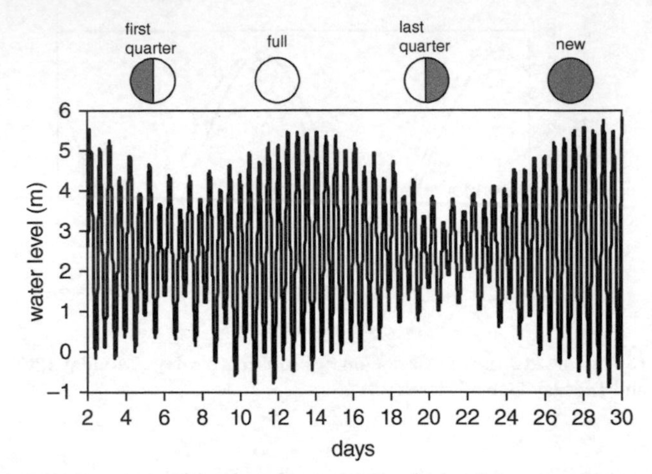

2. The tide at Juneau in March 2017 with the phases of the Moon.

As time advances, the *tidal range*—the vertical distance from high to low water—also goes through a fifteen-day cycle. In Figure 2 we have plotted out the tidal curve at Juneau over twenty-eight days. The tidal range varies from a minimum of less than 3 metres to a maximum of over 6 metres. The times of greatest tides (on the 13th and 28th) are called *spring tides*. On a spring tide, the tide rises to its highest level and also falls to its lowest level. Spring tides occur shortly after the days of new or full moon; at these times, the Sun and Moon are in line with the Earth and their tidal forces combine to create large tides. You can see in Figure 2 that spring tides occur a little after the day of full moon. Most places have an interval of one or two days between new or full moon and spring tides; the lagging of spring tides behind the phase of the Moon is called the *age of the tide* and we return to this in Chapter 2. Between the spring tides are periods of smaller tidal range called *neap tides* which occur at the quarter moons when the Sun and Moon make a right angle with the Earth.

Because the tidal range and the time of high water both have a cycle with the same period of half a month, they are synchronized with each other. If, for example, today is a spring tide and high water is at 10 am, then fifteen days from now it will again be spring tides and high water will be back to 10 am. The time of high water at spring tides at any given place always occurs at about the same time of day, give or take a little. This time is called the *tidal establishment* and, if you know it for a particular place, approximate but useful tidal prediction becomes easy.

For example, at Juneau, high water of spring tides occurs at 1.30 pm local time on the 13th of March. It will occur at about this time on other spring tides: the tidal establishment at Juneau is, therefore, 1.30 pm. On days other than spring tides, the time of high water will depend on how many days have elapsed since the last spring tide. For example, three days after spring tides, high water at Juneau will be 3x50 minutes after 1.30 pm, that is, at about 4 pm. A glance at the Moon and knowledge of the tidal establishment (with, for a professional twist, an adjustment to allow for the age of the tide) is all you need to do this calculation.

Cycles, cycles everywhere

This method of calculating the time of high tides is fine for working out when to take the family to the beach, but it isn't accurate enough to bring ships into harbour safely. There are other harmonics, or rhythms, to the tide. In particular, the fact that the Sun also creates tides changes the period between high waters in a regular way through the month. In the week following a spring tide, the period of the semi-diurnal tide is reduced to below twelve hours and twenty-five minutes and then increased in the following week. This is why we were a bit vague about the exact interval between high tides in the previous paragraphs. The discrepancy accumulates such that the error in assuming a fifty-minute interval in the time of the tide from one day to the next is greatest at neap tides, or quarter moons. William Hutchinson, the harbourmaster at Liverpool in England in the second half of the 18th century, wrote

> I have observed ships coming in at neap tides, about the quarters of the moon, when instead of meeting with high water as expected by the common way of reckoning...they have often struck or come aground.

The problem was solved by the production of accurate tide tables that added the effects of the Moon and Sun tides and allowed for all the perturbations caused by the rather complicated motion of the Earth, Moon, and Sun.

Daily and longer rhythms

On most days in Figure 2 the two high waters and two low waters are not at the same level. This is particularly apparent at neap tides on, say, the 20th when there is a difference of up to a metre between the two low waters on the same day. The difference in the level of the two tides on the same day is called the *diurnal inequality*.

This inequality is created by a variation in water level with a daily, or diurnal, period of about one lunar day superimposed on the semi-diurnal tide. To take an example, if at a particular time the high water in the *diurnal tide* coincides with high water in the semi-diurnal tide, it will raise the level of that high tide. Twelve and a half hours later, the low water in the diurnal tide will coincide with the next high in the semi-diurnal tide and that will be dragged down, creating a difference in the level of two consecutive high waters. The same principle applies to two consecutive low waters. We can often see the effects of the diurnal inequality in high tides on a beach when there are two parallel high water marks—strandlines left by succeeding tides (Figure 3).

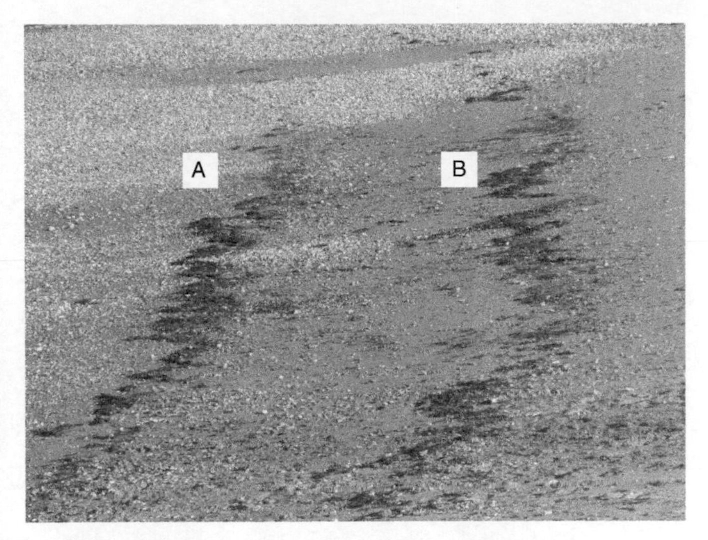

3. Strandlines left by successive high tides on a beach in Pembrokeshire, Wales. Strandline B is higher up the beach than A because of the diurnal inequality of the tide.

In most parts of the world, the diurnal part of the tide is small compared to its semi-diurnal companion, but it is always there in the background. In just a few places it breaks through and the main tidal rhythm becomes diurnal and there is just one high water in the day. This happens on the north-east coast of Australia and in parts of the west coast of North America. The relative importance of the diurnal tide is sometimes quantified by a form factor calculated as the ratio of the size of the diurnal and semi-diurnal tides.

The diurnal tide is produced by the fact that the orbit of the Moon (and that of the Sun) is inclined to the plane of the Earth's equator (we have sketched this in Figure 7 in Chapter 2). As the Moon orbits the Earth it spends one half of the month north and the other half south of the equator. The diurnal tides are largest when the lunar declination (the angle the Moon makes with the plane of the equator) is greatest. Diurnal tides disappear twice each month when the Moon crosses the equatorial plane. Conversely, the semi-diurnal tide is largest when the declination of the Moon (and Sun) is small.

The largest semi-diurnal tides of the year occur at the equinoxes, in March and September, when the solar declination is zero. If the lunar declination is also small at these times, extra large *equinoctial tides* result. An additional factor is that the Moon's orbit is elliptical: its distance from the Earth varies during the month. Equinoctial tides are greater still if they coincide with lunar perigee: the time of the Moon's closest approach. As the years advance these cycles of declination and distance move in and out of phase producing a rhythmic pattern of large tides with a period of several years.

In Box 1 we note some of the places with exceptional tides. The Bay of Fundy in Canada is usually acknowledged to have the

Box 1 Highest and fastest

The United States National Oceanographic and Atmospheric Administration (NOAA) lists the places where the greatest tides have been observed. They use the *mean* tidal range in compiling their list; the range will be greater at all places on a spring tide. Top of the list is the Bay of Fundy in Canada. At Burntcoat Head in the Minas Basin at the upper end of the Bay of Fundy, the mean tidal range is 11.7 metres. This figure rises to 12.9 metres on an average spring tide. The vertical rise in water level from low to high tide is equivalent to the height of a three-storey house, complete with roof and chimneys.

A number of places vie for second place in the table. These are Ungava Bay in Hudson Strait, Canada (9.76 metres), Avonmouth, UK (9.60 metres), Cook Inlet, Alaska (9.23 metres), Rio Gallegos,

Argentina (8.84 metres), and Granville, France (8.60 metres). There is no obvious pattern to the geographical distribution of these places, other than that the coast, rather than mid-ocean, is the place to look for large tides. The important thing about them is that they lie in seas which resonate with a regularly applied tidal force, as we shall see in Chapter 4.

Fast tidal currents are observed where, each tide, a large volume of water is forced through a narrow sea strait. This happens when, for example, the tide in the open sea forces a tide in an inland sea through a narrow strait. The Saltstraumen in Norway is a sea strait which experiences some of the fastest tidal currents in the world. In a semi-diurnal tide the currents reach their maximum speed four times a day. In the Saltstraumen, the currents peak at several metres per second and create strong whirlpools or maelstroms. Other areas of particularly fast tidal currents (and so of interest for tidal power) are Pentland Firth in Scotland and Cook Strait in New Zealand.

greatest tidal range but it is somewhat unusual because the change from a neap to spring tide in the bay is not great. Although Fundy undoubtedly has the largest *mean* tidal range, on days with a spring tide Ungava Bay may have a larger range.

Life between tides

The seashore marks the transition between the marine and terrestrial worlds and it is rarely clear-cut. Water level varies with the tide, wave action, and weather effects, and the area of shore directly above the water level is wetted by wave spray. The result is a strip of coastline, neither fully marine nor fully terrestrial, with strong environmental gradients. The vast majority of organisms living here are marine and, to them, the duration of emersion (exposure to air) matters. For a tidal shore, the emersion pattern varies with the tide and distance up-shore. A higher position will be emersed for longer (between successive high tides, or perhaps even successive high waters of spring tides). Periods of emersion create stresses: desiccation; thermal shock; strong irradiation; lowered salinity; and reduced time with which to find food or reproduce. Organisms have adapted in various ways to tolerate these stresses and this often creates the patterns of vertical zonation, the horizontal banding of organisms on rocky shores or cliff faces familiar to many of us.

Zonation patterns are most distinctive in sessile organisms (those attached to the seabed), but also occur in mobile animals. Physical stresses tend to set the upper boundaries of zones, whereas biological factors (e.g. competition and predation) influence the lower boundaries. Zonation occurs worldwide and the patterns are remarkably similar for a given type of coast. They are most striking on macrotidal rocky shores, where the tidal range and hard substrate combine to give a broad marine–terrestrial transition with clear zones in the macrofauna and -flora. Zonation occurs in other environments, for example mudflats and sandy beaches (where the substrate is too unstable to permit the attachment of larger organisms, but some digging and sieving of sediments reveals zonation of the infauna).

The intertidal environment is among the most challenging to inhabit. Physical variables may not achieve the values in those environments deemed 'extreme' (e.g. the poles, deserts, hydrothermal vents) inhabited by so-called extremophiles, but the wide, rapid (almost instantaneous), and frequent fluctuations of physical variables *are* extreme. There is a bewildering array of adaptations to life under these conditions, achieved through long-term evolution. Some species of limpet always return from foraging to tightly-fitting home 'scars' etched into the rock to minimize desiccation during daytime low tides and to reduce the risk of predation. Shells (e.g. mussels) can be clamped shut, plates can be pulled together (e.g. barnacles), and opercula can be drawn across apertures (e.g. periwinkles). There are species that crowd together to create microhabitats that retain moisture, and algae that tolerate up to 90 per cent water loss, drying out to the point of brittleness only to rehydrate for 'business as usual' on the next tide. Then there are the burrowers, the rock-borers, those that produce slimy protective coatings, and a range of organisms that make metabolic adjustments (e.g. to conserve water and prevent the build-up of hazardous waste products). Many of the more mobile intertidal animals seek refuge under seaweeds, in rock pools and moist crevices during low tide, often finding their way by a complex combination of responses to various physical stimuli. Rock pools are, as we know, great places to look out for crabs and starfish.

To some intertidal animals, there are advantages in anticipating the tides or retaining a 'tidal memory'. Many of these creatures exhibit cycles of behaviour with semi-diurnal and spring–neap tidal frequencies that continue for some days, in phase with the tide from their shore, when the animal is removed from the shore and placed in constant laboratory conditions. Classic examples include the shore crab *Carcinus maenas*, the sea louse *Eurydice pulchra*, and the flatworm *Symsagittifera roscoffensis*. As Ernest Naylor shows in his book *Moonstruck: How Lunar Cycles Affect Life*, there is growing evidence that these 'circatidal' and 'circasemilunar' rhythms, like the more familiar circadian rhythms, are controlled by internal biological clocks, which have evolved to give an adaptive advantage in the intertidal zone and which have a molecular, genetic basis (so-called 'clock genes'). The associated scientific field of chronobiology is not for the faint-hearted: deniers of these notions would have it that residual variables, unaccounted for in the laboratory experiments, such as barometric pressure, the Earth's magnetic field, the Moon's gravity, and even cosmic radiation, could somehow be providing directly-sensed stimuli that perpetuate such behaviour without the need to invoke endogenous rhythms. Why then do the cycles break down after days or weeks in the laboratory? And why can they be re-instated and manipulated by simulated tides? It seems tides provide external environmental synchronization, and these animals require repeated exposure to help their biological clockwork 'keep time'. There are undoubtedly exciting discoveries still to be made in this area.

Meteorological effects

Figure 4 shows a water level record from New York in which the observed and predicted water levels are compared. The two curves don't always agree. The predictions are good at getting the times of high and low tide correct but the water levels are not always right. This is particularly so on the afternoon of the 14th of March when the observed high tide is almost 1 metre higher than predicted. These higher-than-predicted water levels are caused by meteorological effects and are called surges. Surges can be produced by strong onshore winds piling the sea up on a shore or by locally low atmospheric pressure. (A fall in atmospheric pressure of 1 millibar leads to a rise in sea level of roughly 1 centimetre and is called the inverted barometer effect.) Surges

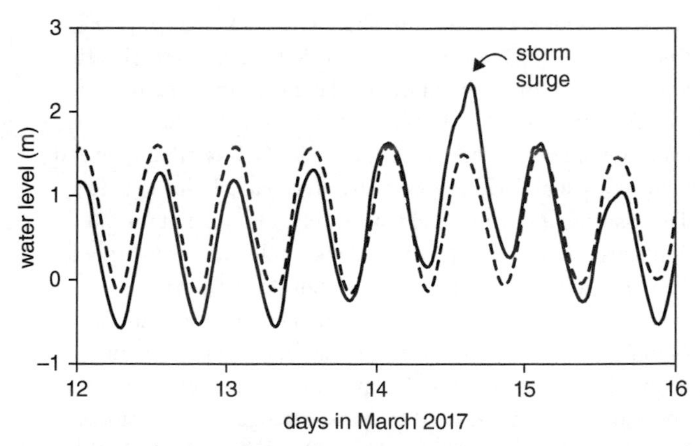

4. A small storm surge in New York harbour. The predicted tide is shown as the dashed line and the observed tide as the continuous line.

can cause serious flooding, especially when they coincide with a high spring tide.

Flooding of low-lying coastal land by *storm surges* on top of a high tide can claim lives. One of the most vulnerable coasts is that of Bangladesh, which has suffered a number of devastating storm surges. One in 1970 killed an estimated half a million people. In Europe, a surge on the night of 31st January to 1st February 1953 killed over 2,000 people living close to the shore of the southern North Sea. It was in response to this disaster that the barriers were constructed to protect London and Rotterdam from future storm surges.

Chapter 2
Making tides

Early maritime civilizations knew that the Moon was important for making tides but a plausible explanation of how this worked, exactly, was slow in coming. An early Arabic idea was that the tide was caused by the thermal expansion of seawater warmed by moonlight. In the 16th century in Europe, serious thought was given to the possibility that the tide was relevant to the new ideas placing the Sun at the centre of the solar system. Galileo Galilei (1564–1642) developed a theory which would allow tides to be created by the movement of the Earth about its own axis and about the Sun. Johannes Kepler (1571–1642) appears to be one of the first to suggest that the Moon is able to attract the waters of the ocean, although this idea on its own is not enough to explain why there are two tides each day.

In 1687, Isaac Newton presented, in the *Principia*, the inverse square law of gravity which, together with his three laws of motion, provided—for the first time—a coherent explanation of the movement of the planets and moons of the solar system. In modern language, the Law of Universal Gravitation can be stated:

> Every object in the universe attracts all other objects with a force directly proportional to the product of their masses and inversely proportional to the square of their distance apart.

Newton's ideas met some resistance at the time (how can objects exert a force on each other when there is no apparent link between them?)—but they explained the workings of the solar system so well that they soon became accepted. Newton devoted some of the *Principia* to an explanation of the ocean tide. The theory didn't help immediately with the practical problem of predicting tides but it did explain many features that had been difficult to account for previously, including the fact that there are two tides a day. Newton also showed how the interaction of the Sun and the Moon, and the lunar declination are important to the tide.

The tide-generating force

The trick to understanding how the Moon raises tides in the ocean is to appreciate that the Earth moves, once a month, around its common centre of gravity with the Moon. This is not an easy concept to grasp; intuition tells us that the Earth remains still while the Moon moves around us. But, it is the nature of bodies held together by gravity that they orbit *about each other* in such a way that their mutual gravitational attraction provides the necessary centripetal force (or if you prefer, balances the centrifugal force) arising from the motion.

In the case of the Earth and the Moon, the common centre of gravity—the barycentre—lies much closer to the centre of the Earth than to that of the Moon; in fact it lies within the body of the Earth about three-quarters of the way from the Earth's centre to its surface. Once a month, the centre of the Earth completes a circle about the barycentre keeping a position such that the Moon always lies on the opposite side of the barycentre. For the next few paragraphs, forget that the Earth is rotating about its own axis and concentrate on this monthly motion of the Earth, which we have sketched in Figure 5.

This figure shows the centre of the Earth, *C*, moving in a circle about the barycentre, *B*. We have drawn a line from *C* to an

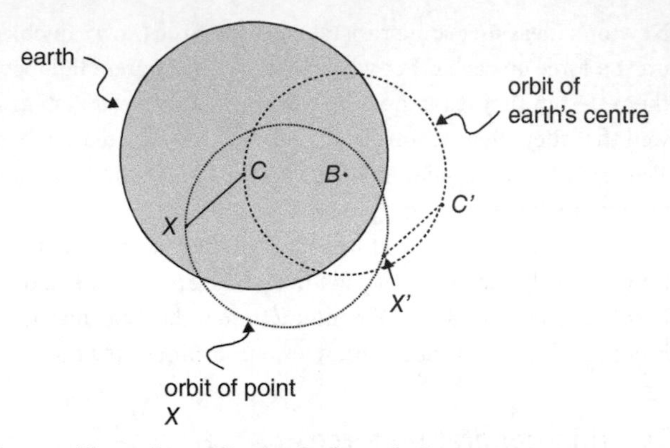

5. The monthly motion of the Earth about the barycentre.

arbitrary point, labelled *X*, within the Earth. Because the Earth is not rotating, the line *XC* keeps the same orientation during the month. We can imagine that *X* and *C* are joined by a solid bar which always points in the same direction. As *C* moves round its orbit it drags *X* with it: for example, when *C* has moved to position *C'*, *X* has moved to *X'*. The point *X* goes around the same sized circle as *C*; because the position of *X* is arbitrary, it follows that all points on and in the Earth go around the same sized circle once a month.

Centrifugal force depends just on the radius of the circular motion and the time taken to complete the circle. Since each part of the Earth moves in the same sized circle in one month, the centrifugal force is the same everywhere. It is directed away from the Moon, parallel to a line connecting the Earth and Moon centres.

The gravitational pull of the Moon, however, varies over the Earth, becoming weaker with increasing distance from the Moon. The Moon's gravity and the centrifugal force balance exactly at the centre of the Earth but at other points there is an imbalance: a residual force acting towards the Moon in the Earth hemisphere closest to the Moon and away from the Moon in the opposite

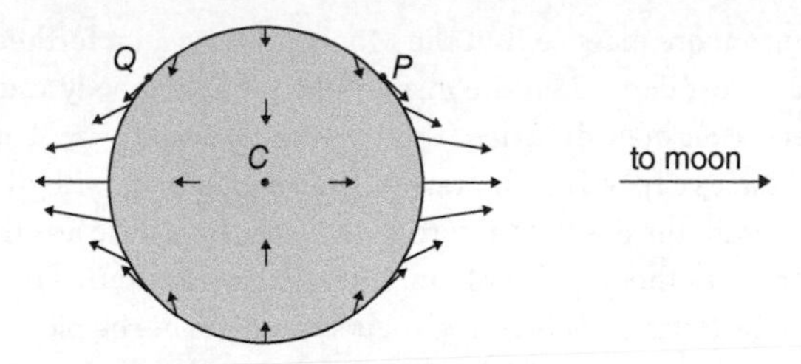

6. The tidal forces created on the surface of the Earth by the Moon.

hemisphere (see Figure 6). At *P*, for example, the Moon's gravity is a little stronger than the centrifugal force and at *Q* it is not quite so strong.

These residual forces are the tidal forces of the Moon on the surface of the Earth; at each point the tidal force is equal to the difference between the Moon's gravity at that point and the Moon's gravity at the centre of the Earth. The maximum size of the Moon's tidal force is very small—about one-ten-millionth of Earth's own gravity at its surface. It seems incredible that such a small force can create the large tides we observe in the ocean. But, as we shall see, the ocean is very good at responding to a small force applied regularly at just the right interval of time.

The tidal forces can be divided into a component acting perpendicularly to the Earth's surface (which adds or subtracts a little from the Earth's own gravity) and a component that acts parallel to the surface. It is this latter component that moves the ocean and creates tides. This horizontal component is called the *tide-generating force*, or the tractive force. As the Earth spins on its axis once a day, these tractive forces act first one way and then another, creating an oscillating force acting on each piece of the Earth and its ocean.

The orbit of the Earth about the Sun generates an equivalent set of tide-generating forces converging on points directly below the Sun and on the far side of the Earth to the Sun. Although the Sun is much more massive that the Moon, it is also a lot further away. Tidal forces depend on the mass of the attracting body and the inverse *cube* of its distance (this inverse cube rule arises because tidal forces depend on the rate of change of gravity with distance). As a result, the Sun's tidal forces on Earth are a little less than half as strong as those of the Moon. The tidal forces exerted on the Earth by other planets in the solar system are negligible.

Despite its smallness, it is possible to measure the tide-generating force. A pendulum bob suspended on a string 10 metres long will be pulled aside by less than one-thousandth of a millimetre by the Moon's tractive force. Observed deflections of a carefully designed pendulum are actually somewhat smaller than those predicted theoretically, by a factor of about 0.7. This reduction is caused by the response of the solid Earth to the tidal force. The earth changes its shape (and the surface tilts) under the influence of tidal forces, an effect known as the Earth tide. As the surface tilts, a component of Earth's gravity acts parallel to the surface and is able to cancel out, in part, the tide-generating force.

The equilibrium tide

If the Earth was covered completely with ocean, the Moon's tide-generating force would cause the water to flow to a point directly below the Moon and another point on the far side of the Earth from the Moon. Bulges in the ocean surface would be created at these places. Newton anticipated intuitively the shape of such an ocean and, in 1740, the Swiss mathematician Daniel Bernoulli derived the exact shape mathematically. The shape is that of an ellipse rotated about its long axis: roughly the shape of an egg (see Figure 7). The sloping water surface creates a horizontal pressure gradient which balances the tide-generating force: the ocean comes into *equilibrium* with the tidal force applied to it. The size of the bulges required to balance the tide-generating force is not great—the ocean is stretched about a quarter of a metre above the no-tide level.

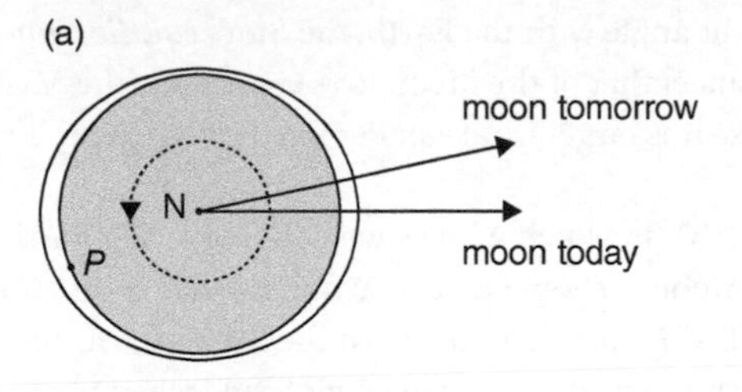

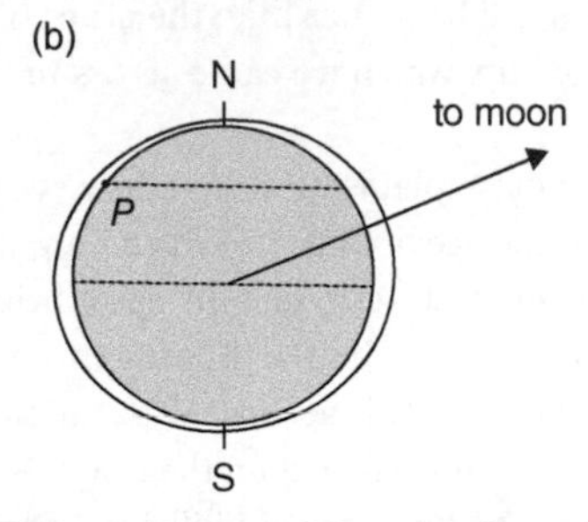

7. The equilibrium tide viewed from (a) above the north pole and (b) above the equator.

If we now start the Earth spinning about its own axis (Figure 7(a)) and *the ocean maintains its position relative to the Moon*, there will be two high tides each day as a point *P* on the Earth's surface passes through the tidal bulges in the ocean. Furthermore, during the time it takes the Earth to turn once on its axis, the Moon moves about one-thirtieth of the way about its orbit. The Earth takes a further one-thirtieth of a day, or fifty minutes, to catch up with the Moon. There are therefore two high tides in a period of twenty-four hours and fifty minutes—or one tide every twelve hours and twenty-five minutes.

The Sun will create additional tidal bulges in the ocean pointing towards and away from the Sun. When the Sun, Moon, and Earth are in line (at times of a new and a full moon), these bulges overlap and create a large *spring* tide. When the Sun and Moon make a right angle with the Earth, the Sun's *equilibrium tide* tends to cancel that of the Moon to some extent (the Moon's wins out because it is larger) and smaller *neap* tides result.

In Figure 7(b) (in which we are now looking at the Earth from the side), the Moon is shown at its maximum declination. For the point labelled *P*, the high tide when the Moon is overhead will be greater than the next (or previous) high tide when the Moon is underfoot. There will be two high tides and two low tides in a day, but one of the high tides will be higher than the other. This effect creates the diurnal inequality which we came across in Chapter 1.

The equilibrium tide model explains, in a qualitative way, many of the observed features of the ocean tide. Unfortunately, it fails in two important respects. First, the range of the equilibrium tide is small: just half a metre or so as the Earth turns through the bulges and the spaces between. This accords with the tidal range at oceanic islands but is much smaller than the tide observed at many coasts. Second, high water in the equilibrium tide always occurs when the Moon crosses the meridian at a particular place. We know that this is not the case: the time of high tide varies from place to place on the same meridian. It is not surprising that the theory fails (Newton acknowledged its weaknesses and knew what needed to be done to correct them), but it is a shame that it fails so spectacularly on these particular points which are the most important for tidal prediction.

The response of the oceans to the tide-generating force

Every point in the ocean experiences a tide-generating force changing in strength and direction with rhythms set by the motions of the Earth, Moon, and Sun. In general, this force will have east–west and north–south components, and each of these will change in different ways with time. These forces act on ocean basins which are connected to each other and bounded by continents. Each ocean will respond to the tide-generating force acting on it and also influence (and be influenced by) what happens in adjacent oceans. It's a complicated business. How will the tide behave in such a world ocean?

An accurate answer to this question had to wait until computers became powerful enough to solve, for the whole globe, the equations that govern the flow of water on a rotating Earth. These equations were first written down by the French mathematician Pierre-Simon Laplace in 1775. They allow Newton's second law of motion (acceleration is proportional to applied force) and the principle of conservation of mass to be applied to the ocean. The solution shows that the tide behaves as a series of waves, with the same periods (or rhythms) as the tidal forcing, sweeping around the outside of the ocean basins in great circular movements. The amplitude of the waves diminishes to zero at points near the centre of the oceans, and the waves travel around these points, mostly in an anti-clockwise sense in the northern hemisphere and in a clockwise sense in the southern hemisphere. This picture of the world ocean tide agrees with modern observations which can be made from Earth-orbiting satellites equipped with instruments (called *altimeters*) that can measure the height of the ocean surface.

To understand how this comes about, we will examine the various processes that lead to this behaviour one at a time until we can build up the whole picture. The tide-generating force varies with several rhythms but the most important of these is that set by the spinning Earth within the Moon's orbit; this force has a period of oscillation of half a lunar day or twelve hours and twenty-five minutes. (To see how this works, imagine we are looking down on the Earth in Figure 6 from above the north pole. Starting at a point directly beneath the Moon and following it as the Earth makes a half turn anti-clockwise, the horizontal part of the tidal force is zero, maximum towards the west at P, zero again, maximum to the east at Q, and returns to zero after half a lunar day.)

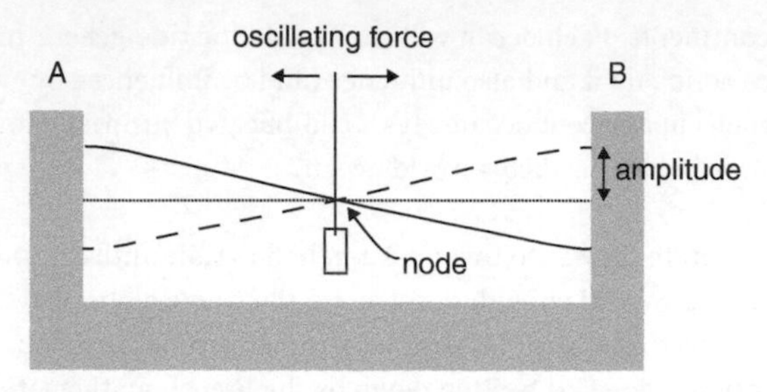

8. A standing wave created in a basin by an oscillating force.

The response of a body of water to an oscillating force can be seen, in miniature, in a kitchen sink. Partly fill a sink or basin, and then move your hand gently back and forth in the water. You will create a sloshing motion called a *standing wave* which will look something like Figure 8 (in fact, it's hard not to create higher frequency oscillations in the sink which complicate the picture, but the simplest motion is the one shown).

The time taken for the water to slosh from one end of the basin to the other and back again is called the period of the wave. Starting at a time when the water level is high at one end, A, the current is everywhere zero. Water then flows from A to B, reaching its maximum speed one quarter of a wave period later; at this time, the surface is flat. One half period after high water at A, it is high water at B and the currents are again everywhere zero. Water now starts to flow back towards A, reaching maximum speed a quarter of a period after high water at B. A further quarter period later it is high water at A, and the cycle is complete.

The water surface oscillates about a line, called a node, across the middle of the basin. At the node, there is no vertical movement of the water surface but the horizontal currents are fastest. Other kinds of standing wave are possible with more nodal lines, but the single node standing wave illustrated in Figure 8 is the simplest and most common.

If you time the movement of your hand just right, you can create large oscillations with little effort: the water can slop right out of the sink. You have found the *resonant period* of your sink. The movement of the wave across the sink proceeds at exactly the same pace as your hand. The force of your hand, acting on the flowing water, continually feeds energy into the wave, which consequently grows. If your hand moves at a different speed to the wave, you will create oscillations in the water which match the speed of your hand, but they will not be so large. The wave now travels ahead of your hand, or gets left behind, and you have to make a new wave. You can feel that your hand has to work harder when you are off the resonant period.

The speed of a water wave in a shallow basin is equal to $\sqrt{(gd)}$ where g is the acceleration due to gravity and d the water depth. The condition for resonance is that one back and forth motion of the applied force (the forcing period) should match the time it takes for the wave to travel backwards and forwards across the basin. The forcing period, T, required for resonance is therefore $T = 2L/\sqrt{(gd)}$ where L is the width of the basin. For a sink 60 centimetres wide with water 15 centimetres deep, the required forcing period to produce a resonant wave is about one second.

Applying the same rule to an ocean basin tells us that, in order to be in resonance with the semi-diurnal tidal forcing, a wave should cross the basin in six lunar hours (a lunar hour is one-twenty-fourth of a lunar day). The speed of a shallow water wave in water 4 kilometres deep (the average depth of the ocean) is 713 kilometres per hour: about the speed of an airliner. Travelling at this speed for six lunar hours, the wave will cover a distance of 4,427 kilometres which is the required basin width for an ocean 4 kilometres deep to be in resonance with the semi-diurnal tide. It's difficult to be sure about ocean widths because coastlines are uneven, but this figure is not so far off the width of major ocean basins such as the North Atlantic. We can therefore expect this, and other oceans, to be close to resonance with the semi-diurnal tidal force.

Laplace's equations can be applied to an ocean-sized basin to see how it will respond, in theory at least, to a small applied oscillating force. This is a familiar problem in classical physics called the forced, damped, harmonic oscillator. To keep things simple, we will just consider one dimension (east to west, let us say): the force acts forwards and backwards along the width of the ocean in the same way as the hand in the kitchen basin. Figure 9 shows the response of an ocean 4 kilometres deep and of varying widths to an oscillating tide-generating force of maximum strength one-ten-millionth of Earth's gravity and period twelve hours and twenty-five minutes. The amplitude of the tide in the ocean rises sharply as the width approaches the resonant width and the *tide wave* becomes most efficient at taking energy from the tidal force. Theoretically, the amplitude grows infinitely large at resonance, but in reality friction prevents this from happening. We have

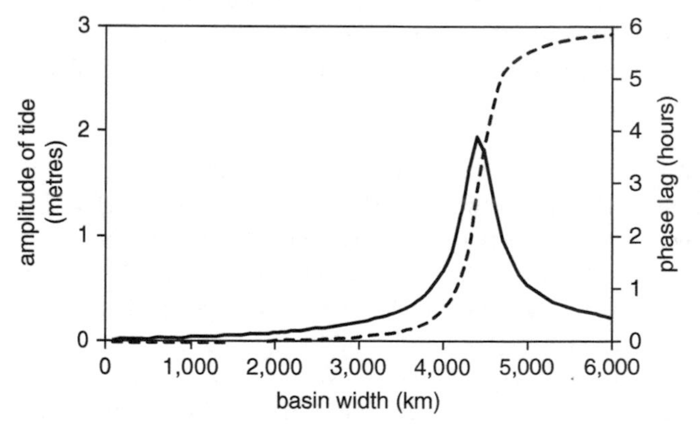

9. The response of an ocean basin 4 kilometres deep and of different widths to the Moon's tide-generating force. The solid curve shows the amplitude of the tide and the dashed curve the phase lag, expressed as the time of maximum elevation at the coast after the time of maximum force directed towards that coast.

included a small frictional damping force in the equations used to produce this figure; more about this later.

Seeing a resonant response on a graph doesn't really tell us exactly what is happening at resonance. How can the tides in the ocean grow so large when the applied force is so small? If you struggle with this concept, try pushing a child on a swing. A small push, applied at just the right regular interval, results in a large swinging motion and a happy child. We find the resonant period of the swing instinctively, to save ourselves from too much work. The applied force of our arms is required only to overcome the friction. The motion actually drives itself, in the same way that a frictionless pendulum, once released, will oscillate forever.

There are three forces acting on the basin: the tide-generating force, the *pressure gradient force* (see Box 2) and friction. The sum of these three forces, according to Newton's second law of motion, gives the acceleration of a unit mass (1 kilogram, for example) of water in the basin.

For a small basin in which the wave crosses too quickly to remain in synchronization with the forcing, the wave, the acceleration,

Box 2 The pressure gradient force

Water pressure at a point below the ocean surface depends on the height of water above that point. When a water surface slopes the pressure is greater under the high part of the slope than it is under the low part. The difference in pressure creates a horizontal force, the pressure gradient force, which acts to drive water down-slope. Alternatively, in steady state, the pressure gradient force can balance other horizontal forces such as the tide-generating force. The pressure gradient force acting on 1 kilogram of water is equal to the gradient of the surface times Earth's gravity.

and friction force are relatively small. The tide-generating force mostly balances the pressure gradient force. High water at the eastern end of the basin occurs at the same time as the maximum tide-generating force in that direction. The size of the tide can be calculated by matching the pressure gradient force to the tidal force. A tidal force one-ten-millionth of gravity will create a surface slope of one in ten million, or 1 centimetre in 100 kilometres. A basin 100 kilometres across and 4 kilometres deep would have tides of just 1 centimetre. This is the kind of tide we could expect in a large deep lake. Tides have been measured in many lakes, but they are always small.

As the size of the basin increases and resonance is approached, the tide becomes bigger, and large accelerations and slopes in the water surface are created. The pressure gradient force now drives the acceleration of the water and the tide-generating force is required only to overcome the frictional drag on the flow. The maximum velocity from west to east in the basin coincides with the maximum tidal force in that direction, and high tide at the east end of the basin occurs one quarter of a cycle, or three lunar hours, after this time.

As the width of the basin continues to increase beyond the resonant width, the slopes and accelerations decrease, and the tide-generating force provides the acceleration. At the time of maximum west-to-east tidal force, the velocities are zero and the water surface slopes down from west to east with low tide at the eastern end of the basin. This is exactly the opposite situation to that in a basin which is narrower than the resonant width.

The effect of Earth rotation

The motion of the tide wave in the real ocean is further complicated by the effect of Earth rotation. Moving objects on a spinning Earth appear to be deflected—to the right in the northern hemisphere and to the left in the south. This effect is called the *Coriolis effect*, after the French mathematician Gaspard-Gustave Coriolis. There is a minimum scale for processes to be affected by Earth rotation: the motion has to continue in the same direction for a significant part of one day. Tide waves, in which the current flows in the same direction for six hours, are affected, whereas the water in a sink is not. Sadly, despite popular belief, the motion of the water down a plughole is not affected by the rotation of the Earth. Instead, the direction of spin down the plughole depends on the residual angular momentum the water retains after the sink was filled. This, in turn, depends on how the taps were used. You can prove this by filling a sink with one tap, leaving the water to settle for a while and then pulling the plug. Repeat with the other tap and you will see the water turns in the opposite direction.

Figure 10(a) shows a hypothetical enclosed ocean basin (in the northern hemisphere) in which there is a semi-diurnal tidal standing wave. High tide occurs at lunar hour 0 on the south shore and at lunar hour 6 on the north shore. At hour 3, the water is flowing at its maximum speed from south to north. The Coriolis effect deflects this flow to the right, creating a slope upwards from west to east. At this time, it is high tide on the eastern shore and low tide on the western shore. At lunar hour 9, the water is flowing at its maximum speed from north to south. The Coriolis effect now makes the surface tilt such that it is high water on the western shore (and low water on the east coast) at this time. The overall effect is that the high water travels around the outside of the basin, anti-clockwise, taking twelve lunar hours to complete the journey.

We can represent this movement, as we have done in Figure 10(a), by drawing lines marking positions along which high water occurs at the same time. These sweep around the ocean basin, like the

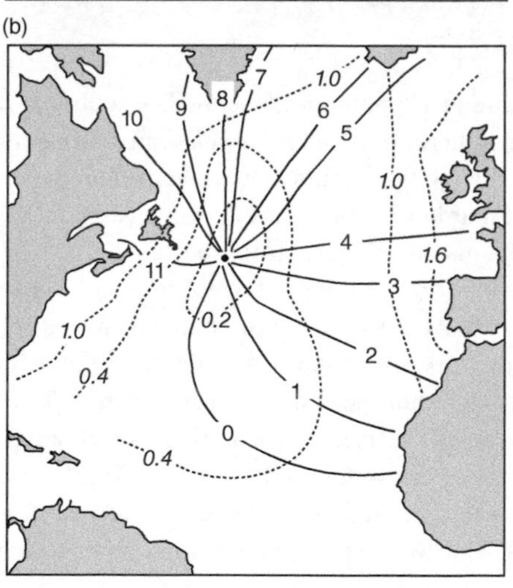

10. (a) A sketch of a forced semi-diurnal tidal wave in a closed ocean basin in the northern hemisphere on a rotating Earth, represented as a co-tidal chart. Continuous lines show the position of the wave crest at different times, in lunar hours. Dashed lines are contours of tidal range, decreasing towards the amphidromic point in the centre. (b) A co-tidal chart of the Moon's tide in the North Atlantic Ocean, with times of high water expressed in lunar hours after the Moon's transit at Greenwich and co-range lines in metres.

hands of a clock, in an anti-clockwise sense in the northern hemisphere. The nodal line pivots about a mid-point as the Coriolis effect tilts the surface first one way and then the other. There is now just a single point of no tide and the amplitude of the tide increases radially outwards from this point, reaching a maximum at the shore. Lines, called co-range lines, can be drawn to represent this and we have sketched a few of these on Figure 10(a).

The tide wave sweeping around a central point of no vertical tide is called an *amphidromic system* and the point of no tide is an amphidromic point. Computer models of the tide show that there are amphidromic systems in all the world's oceans. Generally the movement of the tide wave about the amphidromic point is as expected: clockwise in the southern hemisphere and anti-clockwise in the northern hemisphere. As an example, Figure 10(b) shows a *co-tidal chart* of the North Atlantic which has an amphidromic point located to the east of Newfoundland.

We have made much in this chapter about the importance of resonance to tides in the ocean. It has to be acknowledged, however, that in the main oceans themselves, tides are not, actually, that large; the tide at oceanic islands is rarely more than the half-metre or so we would expect from the equilibrium theory. What is happening is that the near-resonant oceans are leaking energy into the *shelf seas* around their margins. As their name implies, these shelf seas lie on the submerged shelves at the edge of the continents. They are shallow and dissipate tidal energy through friction between tidal flows and the seabed. It is mostly the energy loss in shelf seas which places the frictional damping on tides in the ocean.

The age of the tide

We will return to the effect of friction later in this book. We can say here, however, that tidal friction is the main cause of something mentioned in Chapter 1—that maximum tidal range, or spring tides, occurs a day or two after the time of new or full moon. This 'age of the tide' effect was reported early on (by Pliny the Elder in 77 AD) but it eluded a proper explanation until much later.

To make a spring tide it is necessary for the tides created by the Sun and Moon to coincide, producing a maximum in the tidal range. The Sun's tide-generating force has a slightly higher frequency than the Moon's (it repeats in a period of exactly twelve hours compared to twelve hours and twenty-five minutes for the Moon). If the frequency of the tide-generating force is increased, the tide wave has less time to cross the ocean basin if it is to keep up with the forcing. We could, if we liked, re-draw Figure 9 for an ocean basin of constant width and with the frequency of the oscillating tidal force on the x-axis. The curves would look the same, with a peak in the amplitude of the tide at the resonant frequency of the basin and a phase lag increasing with frequency, most rapidly when the ocean is close to resonance. The exact shape of the curves depends on the frictional damping.

Because the phase lag between the applied force and high tide increases with the frequency of the force, the delay is greater for the Sun's tide than it is for the Moon. When the Earth, Moon, and Sun are in line (at new or full moon) the maximum tidal force of the Sun and Moon occurs at the same time but, because of the different phase lags, the sun's high water is later than that of the Moon. Spring tide cannot therefore occur on the day of a new or full moon.

Let's say that on the day of a new moon, the Sun's high tide occurs one and a quarter hours after that of the Moon. Each day after new moon, the time of the Moon's high water advances by fifty minutes and the time of the Sun's high water stays exactly the same (since it has a period of twelve hours). After a day and a half, the Moon's high water has advanced by one and a quarter hours (1.5 x 50 minutes); the two high waters now coincide and maximum tidal range will occur one and a half days after new moon. The exact delay between new or full moon and spring tides will depend on the phase difference between the Sun and Moon tides at a given location.

Chapter 3
Measurement and prediction

Tides, unusually for natural events, can be forecast with great accuracy for years in advance. The ability to do this and make people safer at sea is probably the greatest practical success story of the science of physical oceanography. There are two aspects of the problem to be considered: the rise and fall of the level of the sea, which we shall call the tide; and the horizontal flow of the water, called tidal streams or currents.

Measuring tides

The tide can be measured most simply with a vertical graduated staff called a tide pole. Tide poles have not been used in recent years as a serious source of tidal information, but they are interesting things to seek out. They can often be found near the entrance to a harbour where they are visible to shipping.
To compare measurements of water level at different places it is necessary to agree where the zero of the tide pole should be placed. This zero level is called the *datum* and different ideas about where it should be prevail in different countries.

One option is to use the lowest level predicted for the tide at that place—the *Lowest Astronomical Tide* (LAT). This is the datum used for measuring depths on British Admiralty charts. In the United States, a datum which is the average height of high or low tides at a place is often used; mean water level is another option. Data can be converted from one form to another, but some knowledge of the local tide is needed to do this. For example, LAT will be half the maximum tidal range below mean sea level.

To make regular measurements of the full rise and fall of the tide for anything longer than a day or so, instruments equipped with a recording device are needed. An early form of tide gauge used a float on the water surface to drive, through a series of mechanical levers, a pen over a rotating roll of chart paper. The unwanted effect of waves on the record was eliminated by placing the float in a vertical cylinder called a stilling well. The top end of the well is open (and above sea level) and the bottom, underwater, end is closed apart from a small hole. Any difference in water level between the outside and inside of the well drives water through the hole, but it takes time for the two levels to become the same. This time, called the response time, varies inversely with the size of the hole. If the stilling well is correctly designed it will have a response time which is long compared to the period of waves but short compared to the period of the tide. Since waves have a period of a few seconds and tides have periods of several hours, a stilling well with a response time of a few minutes works well. The water level inside the well will follow the slow variations of the tide but will not be able to keep up with the more rapid variations in sea level caused by waves.

The principle of the stilling well applies on larger scales too. Apart from in a few special areas, the Mediterranean Sea of Europe has very small tides. It is connected to the Atlantic by the narrow and shallow Strait of Gibraltar. The Strait acts like the hole in the stilling well: the response time of the Mediterranean to fluctuations in Atlantic sea level is measured in months. Tides in the Atlantic move up and down too rapidly to affect the Mediterranean greatly and the tides in the Mediterranean are small.

The trend with modern coastal tide gauges is to avoid contact with seawater altogether so that corrosion is minimized. A popular form of tide gauge mounted on a pier, for example, bounces a short pulse of radio waves off the water surface and determines the distance between itself and the sea surface from the time taken for the wave pulse to make this short journey. For a gauge just a few metres above the sea surface the return time of the radio waves will be measured in nano-seconds, and it is only in recent years that instruments have been sufficiently precise to measure such small times. Sound waves may be used instead of radio, in which case it helps to have a vertical tube to confine the acoustic pulses.

The vertical tide can also be measured by the change in pressure that it causes at the seabed, or at some fixed height above the bed. Pressure gauges have the advantage that they can be deployed in the deep ocean. Changes in pressure are measured by a piezoelectric cell, which creates electric charges when it is squeezed. Pressure gauges are affected by changes in atmospheric pressure: a change in air pressure of 1 millibar is approximately equivalent to a change in water level of 1 centimetre and has to be allowed for. Some pressure gauges are able to measure atmospheric pressure as well as the water pressure, and so a correction can be made.

Measuring tidal currents

The measurement of tidal currents has always been a challenge. For a start, currents are vectors. Two numbers are needed to fully describe a current, for example speed and direction or components of the current in north–south and east–west directions. Second, tidal currents vary with depth. Third, all currents are inherently variable spatially, especially near the coast. Currents measured at one place are not necessarily a good guide even to the direction of the current a short distance away.

A traditional current meter is equipped with an impellor which turns in the flow and a compass to give the direction. The speed and direction of the flow can be recorded on a display and noted down, or stored electronically on a logger.

The instantaneous current at any place can be represented on a map by an arrow. The length of the arrow should be proportional to the speed of the current and it should be drawn to point in the direction of the flow. In the case of a tidal current which changes with time, the tip of the arrow will move around a curve over a tidal cycle. In offshore waters, this curve will often be, approximately at least, elliptical in shape, such as the one marked A in Figure 11. The current reaches a maximum speed when the arrow points along the semi-major axis of the ellipse and a minimum speed three lunar hours later when it points along the semi-minor axis: there is no actual time of zero current, just times of maximum and minimum flow. The sense in which the current vector turns within the ellipse can be either clockwise or anti-clockwise.

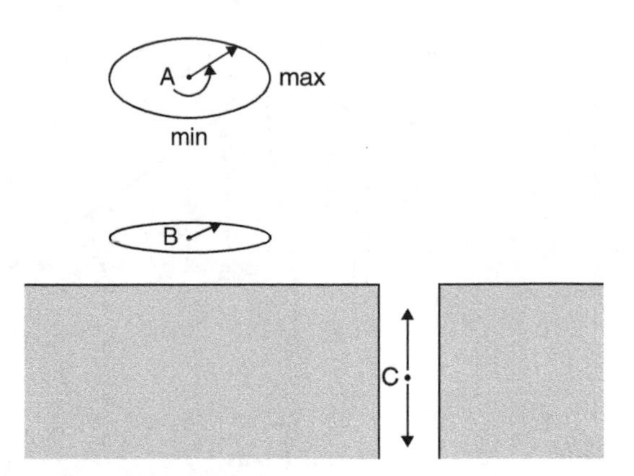

11. Tidal ellipses offshore (A and B) and rectilinear flow in an estuary (C).

As the coast is approached, the land inhibits the flow in a direction perpendicular to the shore and the ellipse becomes flattened (B in Figure 11). The same thing happens in a narrow channel, or estuary (C), and the current will flow in just two directions: either into the estuary (when it is known as the *flood*) or out (the *ebb*). Now there is a time (called *slack water*) when the current goes to zero, as it turns from flood to ebb and vice versa.

The vertical variation of the currents in an estuary can be measured by lowering a current meter to a series of depths below a moored boat, to obtain the profile of the current at regular intervals. An example is shown in Figure 12.

This picture shows two of the characteristic features of tides in estuaries. The surface rises more quickly than it falls and the flood currents are faster than those of the ebb. The relatively fast rising tide is a consequence of the dynamics of flow in shallow water, which we shall return to in Chapter 4; we can note here that the faster flood currents are important for sediment transport.

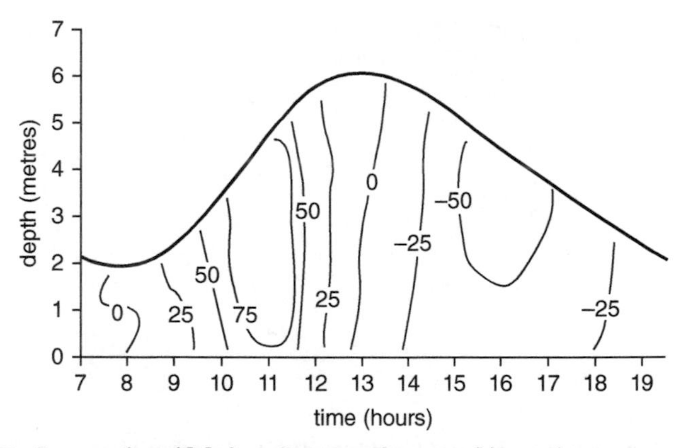

12. Currents in a tidal river. Contours show speed in centimetres per second and the dark line shows the height of the water surface above the bed. Positive values mean that the current is flooding.

More sand and mud is carried into the estuary by the flood tide than is removed by the ebb. Estuaries that are important shipping channels need to be constantly dredged to remove the sediment brought in by the tide.

The vertical variation of tidal currents can also be measured by an Acoustic Doppler Current Profiler (ADCP). This instrument sends 'pings' of sound into the flow. Some of the sound will bounce off particles in the water and travel back to be received by sensors on the ADCP. The time delay between the transmitted ping and the echo gives the distance from which the echo has come. If there is a current, there will be a change in the frequency of the echo (a Doppler shift) and this can be used to work out the speed of the flow. The acoustic pings are transmitted at an angle to the vertical so that they provide information on the horizontal flows. An ADCP can measure the vertical profile of the current more frequently and at a higher vertical resolution than conventional current meters.

Measurements of a current flowing past a fixed point are called *Eulerian* (after Swiss mathematician Leonhard Euler (1707–83)). A completely different way to measure currents is to deploy a tracked drifter that moves with the flow. Such methods are called *Lagrangian*, after Italian mathematician and astronomer Joseph-Louis Lagrange (1736–1813). In one common configuration a surface float equipped with a radio transmitter is attached to a sub-surface drogue. The drogue is designed to catch the flow: a sail or a parachute, for example, works well. The current drags the drogue along and the position of the drifter can be tracked from the radio transmissions of the surface buoy.

For users of British Admiralty charts, tidal diamonds are a familiar source of tidal stream information. A point on the chart at which the currents have been measured is identified by a diamond shape, containing a letter. On the edge of the chart, hourly values of the tidal stream speed and direction at spring and neap tides are tabulated. The time of each measurement is expressed relative to the time of high water at a standard port.

The information needed to make these tables has been gathered over many years. Today the streams are measured by a vessel- or seabed-mounted ADCP. Traditionally, however, the currents were measured by tracking a spar buoy (called a pole logship by surveyors) and the results processed graphically. A spar buoy is a buoyant vertical pole or tube weighted at the bottom so that its upper end just shows above the water surface. The buoy follows the current averaged over its vertical length. It is tracked by a survey vessel (the 'logship') and its position recorded every half hour over a period of either twenty-five or fifty hours. The difference in position of a consecutive pair of measurements gives the speed and direction of the flow.

Historically, the observations were processed manually by plotting them on a polar diagram; that is a piece of paper with equally spaced concentric circles representing flow speed and radial lines marked in degrees for direction of flow. A measurement of speed and direction from the spar buoy plots as a point on this diagram. As more data points are added they form the outline of a tidal current ellipse. The analyst would draw a best-fit curve through these points by eye, averaging out variations between tidal cycles and identifying any 'outlier' observations which may have been adversely affected by wind.

The ellipse is annotated with the time in hours before and after high water at a nearby port. A line drawn from the centre of the polar diagram to each of these hour marks gave the flow speed and direction at that time. The speed is adjusted to give the expected value at mean spring and neap tides. If there is a significant non-tidal flow, the ellipse will be displaced from the centre of the polar diagram. This could be allowed for (and the effect of the residual flow removed) by marking a new central point in the middle of the ellipse.

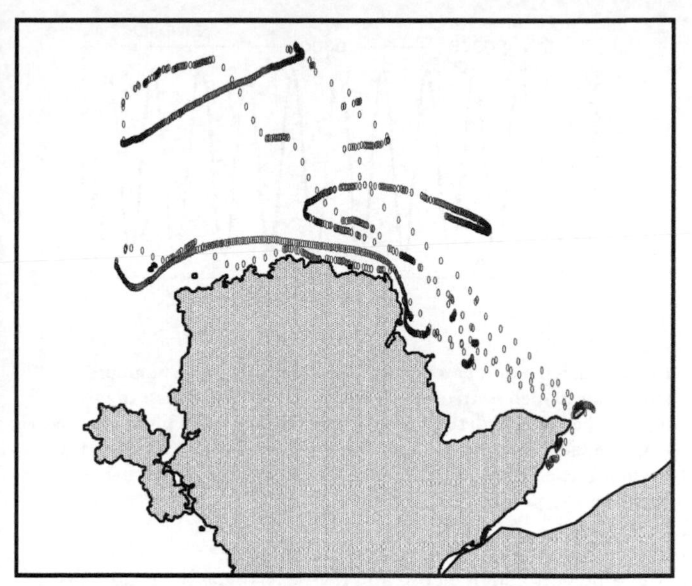

13. Motion of a tagged seabird near Anglesey in Wales. When the bird is flying, the recorded positions are well-spaced but when the bird rests on the water and consecutive positions are close together the points trace out a dark track which shows the tidal current.

New methods of measuring currents continue to be developed. A recent innovation is to use records from tagged seabirds resting on the water. Birds are routinely tagged with small global positioning system (GPS) loggers to study their behaviour. When seabirds rest on the water (usually at night), they become unwitting recorders of the current (Figure 13).

Analysing tidal data

The standard method of analysing tides and tidal currents is a curve-fitting procedure known as harmonic analysis. In the case of tide height, for example, a record of water levels at one place over a period of time is represented by the sum of a number of pure cosine waves, called tidal harmonics (Figure 14). The period

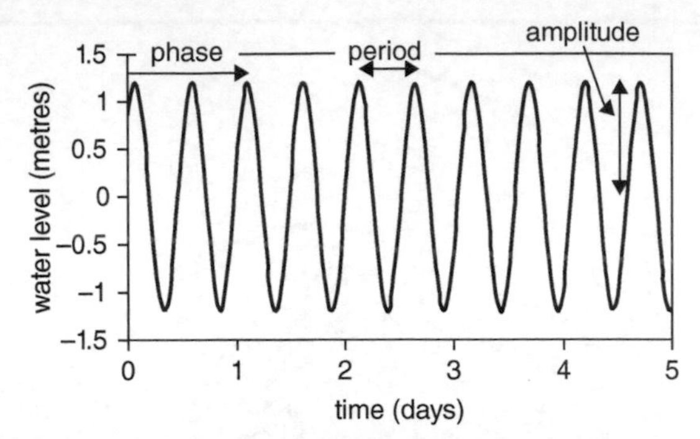

14. A tidal harmonic can be specified exactly by three numbers: amplitude (which controls the height), period (which sets the time interval between high tides), and phase (the time of a high tide relative to a standard time). Once these numbers have been determined, a harmonic can be projected into the future and added to other harmonics to predict the tide.

(the time between one crest and the next) of each harmonic is set by the motion of the Earth, Moon, and Sun, and is the same for all places. The amplitude and phase vary from place to place and are determined by matching the fitted curves to observations.

The most important harmonic is that which would be produced by an idealized Moon in a circular orbit in the plane of the Earth's equator. This harmonic has a period of exactly twelve hours and twenty-five minutes (or 12.42 hours) and is given the symbol M_2 (M stands for Moon; and the subscript 2 refers to the fact that this harmonic has two cycles per day). An idealized Sun, at a fixed distance from Earth and lying in the plane of the equator, produces a harmonic with a period of exactly 12.00 hours which is called S_2.

Adding M_2 and S_2 harmonics together creates a semi-diurnal tide whose amplitude changes over a fortnightly cycle. At spring tides, the two harmonics are in phase and their amplitudes add. At neap tides, they are out of phase and the resultant tide is the difference between the amplitudes of M_2 and S_2. The relatively slow change in amplitude of a signal created by two harmonics close in frequency is called 'beating'. Beating of sound waves can be heard when musicians tune up: two instruments not quite producing the same note create a sound which beats, or fluctuates in volume. Adding M_2 and S_2 harmonics produces a beating of amplitude with period 14.8 days, the period of the spring–neap cycle.

Harmonics with periods of about a day are needed to create the diurnal tides. The two most important of these are called K_1 (with period 23.93 hours) and O_1 (period 25.82 hours). These harmonics have been created just so that, when added together, they beat to produce a semi-monthly variation in the amplitude of the diurnal tide.

There are many other harmonics that can be included to create an increasingly accurate representation of the tide, but we will mention here just one more. The N_2 tidal constituent has a period of 12.66 hours. Combined with M_2 it produces beating with period 27.55 days which is the interval between times of lunar perigee—when the Moon, in its elliptical orbit, gets closest to the Earth.

Let's see how well we can predict the tide at Juneau, Alaska, starting with the observations shown in Chapter 1; the technique can equally well be applied to any location where observations of sea level are available. Of course, the tide at Juneau has already been analysed by the operators of the gauge, NOAA, and so we know what the right answer is. NOAA use thirty-seven harmonics in their analysis, but we will restrict ourselves to just five to keep the maths relatively simple. The amplitude and phase of these five harmonics, as published by NOAA on their website, are shown in Table 1. These are known as the tidal constants for the port.

The *speed* shown in this table is the word used in tidal analysis for the angular frequency of the harmonic. For example, the M_2 harmonic completes a full cycle (of 360 degrees) in 12.42 hours

Table 1. Tidal constants for Juneau, Alaska

Harmonic	NOAA amplitude (metres)	Phase lag (degrees)	Speed (degrees/hour)	Spreadsheet amplitude (metres)
M_2	1.995	282.5	28.984104	2.00
S_2	0.680	315.5	30.00	0.69
N_2	0.405	258.2	28.43973	0.33
O_1	0.311	250.2	13.943035	0.25
K_1	0.520	265.0	15.041069	0.49

and so its speed is 360/12.42 = 28.98 degrees per hour. The phase lag is, by tradition, the time difference between high water in the harmonic and the time at which high water in that harmonic would occur in the equilibrium tide at Greenwich in England. For example, high water in the equilibrium M_2 harmonic at Greenwich occurs when the Moon crosses the meridian at Greenwich. At Juneau, high water in this same harmonic occurs 9.75 hours (282.5° divided by 28.98°/hour) after the Moon crosses the Greenwich meridian.

We can learn some useful facts just by visual inspection of the sea-level record shown in Chapter 1. The tidal range at spring tides is about 5.5 metres and at neaps it is about 2.5 metres. If the M_2 and S_2 harmonics were the only ones that mattered, we would expect the sum of their amplitudes to be half the spring range, or 2.75 metres, and the difference between their amplitudes to be half the neap range, 1.25 metres. This gives us two simultaneous equations for the amplitudes of M_2 and S_2 which can be solved to give 2 metres and 0.75 metres respectively—not so far off the values in Table 1.

Programs to perform tidal analysis can be obtained from international tidal authorities; some programs can be downloaded for free from the internet, but it is never clear how reliable they are. It is also possible to learn the principles of the job in a spreadsheet that can perform a *multiple* regression. The dependent variable is the observed sea level and the independent variables are the harmonics whose amplitudes and phases we wish to determine. The spreadsheet is set up with time in column 1 and the observations of sea level in column 2 (hourly observations are fine). Subsequent columns are filled with sine and cosine terms having the frequency of a chosen harmonic and with unit amplitude. Performing a multiple regression then returns the amplitudes which give the best fit to the observations when summed. The sine and cosine terms could be combined into a single cosine harmonic with a phase lag using a trigonometric relationship, but this is not necessary in order to plot out the predicted tide.

The harmonic amplitudes from the spreadsheet analysis of the Juneau data are shown in Table 1. There are differences from the NOAA estimates but these are explained by the relatively short record of thirty days that we used for the analysis and the fact that we left out many of the harmonics included by NOAA.

The five harmonics add to make a good fit to the observed tide at Juneau as you can see in Figure 15. To make a prediction of the tide, we then choose a day in the future, calculate the time difference between the start of that day and the start of our record,

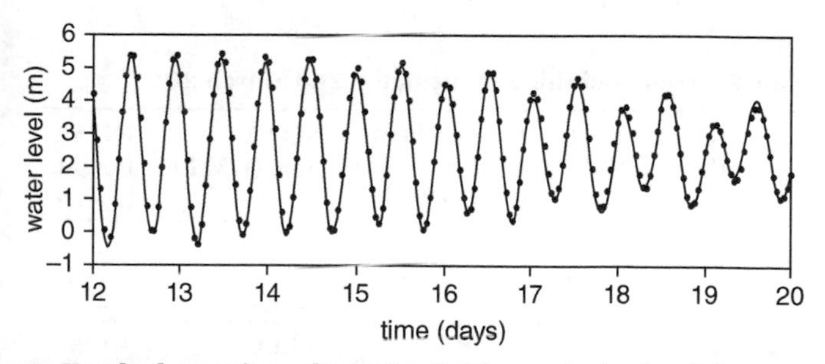

15. Hourly observations of water level at Juneau (points) and the curve fitted by multiple regression on five harmonics.

and work out the values of the harmonics over the twenty-four hours of that day. The harmonics can be summed to give the predicted tidal curve. If the prediction is to be made a long time after the observations, it is important to specify the speed of the harmonics as precisely as possible.

We compare our prediction of the tide with the NOAA predictions exactly one year after the beginning of the observations—2nd March 2018 in Table 2. The method gets the heights of the tide correct to within 10 centimetres and the times to within six minutes. That wouldn't be good enough to set up in business but it does show that we understand how the business is carried out.

In principle, the more harmonics that are added to the tidal analysis, the more exact will be the fit to the observations. There is, however, a limit to the number of harmonics that are meaningful. For a tidal analysis to be able to separate the amplitudes of two harmonics of frequencies f_1 and f_2 we need a record length at least equal to $1/(f_1 - f_2)$. To resolve harmonics that are closer together in frequency requires a long record. No professional analysis would be carried out with a record as short as a month: twelve months would be a minimum requirement.

Harmonic analysis of observations and the prediction of tides are now routinely carried out by computer programs, but before

Table 2. Predicted tide at Juneau for 2nd March 2018

	Spreadsheet time (GMT)	Spreadsheet height (metres)	NOAA time (GMT)	NOAA height (metres)
Low	03:57	-1.00	03:54	-1.05
High	10:13	5.58	10:19	5.54
Low	16:13	-0.18	16:13	-0.23
High	22:17	5.78	22:21	5.87

computers were widely available, other methods were used. Clever ways were devised to analyse tidal records by hand to derive the tidal constants.

The principle of these methods was to divide a record into segments containing a (small) whole number of cycles of a single harmonic. Within each segment, the high and low tides of the selected harmonic would always appear in the same places, but this would not be so of other harmonics. When a large number of segments were averaged, the signal of the chosen harmonic would reinforce itself and the other harmonics would cancel out.

The method is most easily illustrated for S_2 which has a period of 12.00 hours. The tidal height at 00 hours each day is averaged over all 24-hour segments of data (and also at 01 hours, 02 hours, and so on, up to 24 hours if we choose to consider two complete cycles of S_2 at a time). As the number of days of observation increases, the pure S_2 harmonic appears. Figure 16 shows the results of this averaging procedure applied to Juneau.

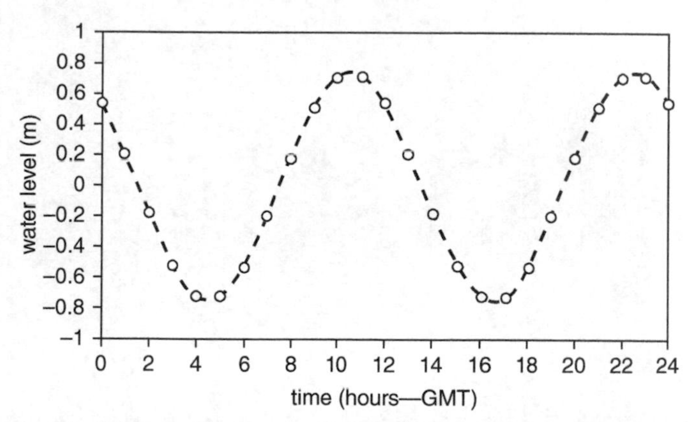

16. The S_2 harmonic at Juneau, Alaska. The points show average elevations at hourly intervals during the day calculated using many 24-hour segments of data (see text). The dashed curve is a cosine curve fitted to the points.

The amplitude of the S_2 tide that emerges is 0.74 metres and the phase is such that high tide in the S_2 harmonic occurs at 10.30 (am and pm) Greenwich Mean Time (GMT). The amplitude is a little larger than that given by NOAA but the phase is exactly right (since from Table 1, for S_2, dividing the phase lag of 315.5 degrees by a speed of 30 degrees per hour gives the time of high tide as 10.5 hours GMT).

In pre-computer days, once the harmonics for a particular port had been determined, the tide would be predicted using a machine. These tide-predicting machines, often made of brass

17. A tide-predicting machine designed by Arthur Doodson in the late 1940s and now on display at the National Oceanography Centre, Liverpool.

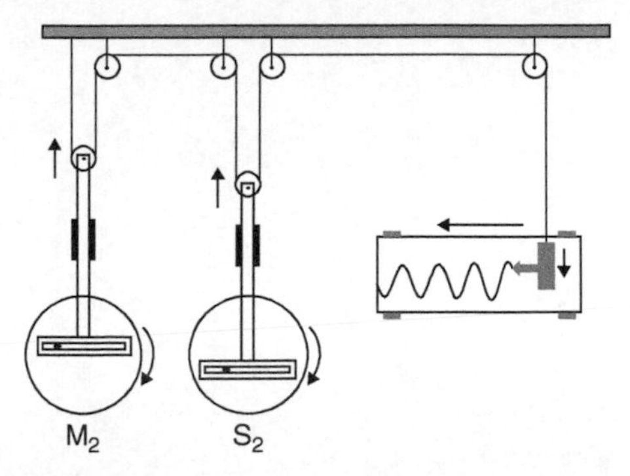

18. Principle of operation of a tide-predicting machine, illustrated with two harmonics.

and mahogany, were beautiful examples of scientific instrument making. The original machines were designed by William Thomson (Lord Kelvin), and by the 1930s all of the major sea-going countries had their own machines. Many have been preserved in maritime museums around the world and are worth looking out for (see Figure 17).

The principle of operation is illustrated in Figure 18. Each of the tidal harmonics used in the prediction is represented by a wheel which turns at a speed proportional to the angular frequency of the harmonic (this may well have been how the word speed came to be used for frequency in tidal work). For example, the wheel representing the S_2 harmonic turns once as the chart paper advances the equivalent of twelve hours. The motion of the wheel is transferred to the pen through a piston driven by a simple cam mechanism. The position of the pin which drives this cam is set to represent the amplitude and phase lag of the harmonic at the chosen port. In Figure 18 we illustrate the procedure with just two harmonics: M_2 and S_2; more wheels are added to represent further harmonics.

Chapter 4
The tide in shelf seas

The great ocean basins are bordered by shallow seas lying on the *continental shelves*: extensions of the continents which flooded at the end of the last ice age. Shelf seas are generally less than 200 metres deep and vary in width from almost nothing to hundreds of kilometres. It is in these shallow seas and the rivers that flow into them that the most spectacular tides are found.

Progressive waves

As the tide rises and falls in the ocean it sends waves, with tidal period, travelling across the continental shelves towards the shore. The tidal forcing of the shelf seas by the ocean tide is more important than the direct action of the tide-generating force. If, for example, the tide in the ocean rises by half a metre at the edge of a continental shelf 100 kilometres wide, the pressure gradient force created by the tilted water surface is 0.5/100,000 or 5×10^{-6} times that of Earth's gravity, fifty times larger than the Moon's tidal force.

Water waves which travel, called *progressive waves*, consist of a series of moving crests (or high waters) interspersed with troughs (or low waters—see Figure 19). The speed at which the crest of a tide wave travels across the ground is equal to $\sqrt{(gd)}$ where g is the acceleration due to gravity and d the water depth (this formula

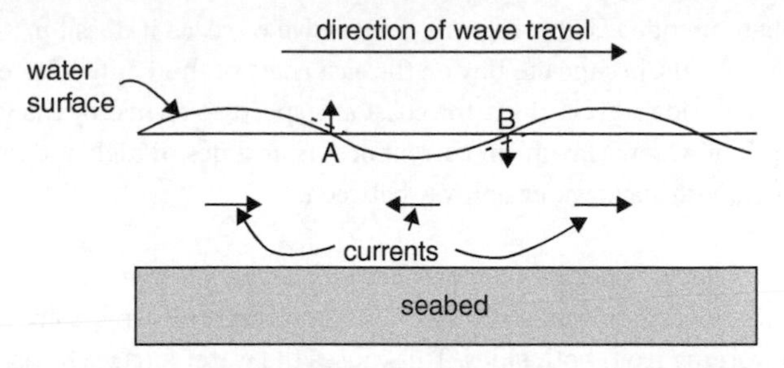

19. A progressive wave in shallow water showing how the currents move beneath the wave.

works so long as the waves are in shallow water compared to their wavelength; this is always true for tide waves). Tide waves travel on the shelf more slowly than they do in the deep ocean but they still move quickly. In water of depth 100 metres, for example, the wave speed is 113 kilometres per hour. Progressive tide waves in shelf seas move at the speed of a fast car on a good road.

The waves are also long. The wavelength—the distance between crests—is equal to the wave speed times the period. For an M_2 tide wave with period 12.42 hours travelling in water 100 metres deep, the length of the wave is 1,400 kilometres. Only a fraction of the wave will fit into most of the world's shelf seas.

Beneath the sloping water surface, the pressure gradient force creates accelerations. Between a low water and the following high, the acceleration is in the direction of wave travel and it accumulates such that maximum velocity in this direction occurs beneath the crest of the wave. Maximum velocity in the opposite direction occurs beneath the wave trough. As the wave travels past a fixed point, the current oscillates, flowing first with and then against the direction of wave travel. You can see this by throwing a small stick onto waves passing the end of a pier: the stick moves backwards and forwards as the waves pass through.

When the tide is behaving as a progressive wave, as it does for example in Chesapeake Bay on the east coast of the United States, the high tide travels along the coast at a speed governed by the depth of water. Maximum current occurs at times of high and low water, with slack water halfway between.

Water waves progress, or move forward, in the following way. In the column of water at point A in Figure 19 the currents are converging from both sides. This causes the water surface to rise at A and this has the effect of making the sloping face of the wave move forward to the position shown by the dashed line. At point B on the opposite face of the wave, the currents are diverging. As water flows away from B, the surface falls and moves forward to the dashed line. The effect is that the *shape* of the wave moves along while the water (as we have just seen) oscillates about a fixed point.

The speed of the currents in a progressive wave depends on the height of the wave, the water depth, and the acceleration due to gravity. Typically, tidal currents in shallow water are of order 1 metre per second. They can be faster in extreme cases but will rarely exceed 10 metres per second. Generally, the speed of the current in a progressive tide wave is much less than the speed at which the wave travels, but this can change as the wave enters very shallow water—in a river, for example.

Standing waves

When a progressive wave hits a coast it bounces back (coastlines appear to be very good at reflecting long tide waves). The tide near the coast behaves as the sum of the incoming and reflected waves, creating a pattern called a standing wave (see Chapter 2). The water surface oscillates between the positions shown by the solid and dashed lines in Figure 20.

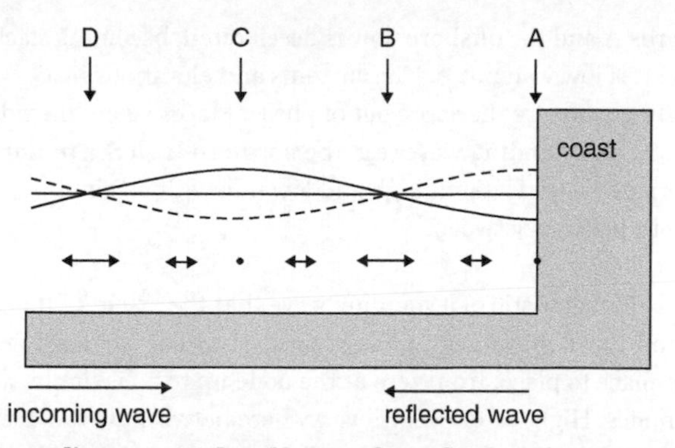

20. A standing wave produced by a perfect reflection of a progressive wave at a coastline. A and C are antinodes and B and D are nodes. Double-headed arrows represent the size of the currents oscillating back and forth beneath the wave.

Standing waves can often be seen when sea waves reflect off a harbour wall: the water goes up and down without the wave going anywhere. The pattern is created by the interference of two equal waves travelling in opposite directions. There are places, called *antinodes* (A and C in Figure 20), where the waves always add constructively and the surface goes up and down the most. Between these there are *nodes* (B and D) where there is no vertical motion of the water surface. At the nodes, the elevation of the incoming wave cancels that of the reflected wave at all times. For this to work, the reflected wave must always be half a wavelength out of phase with the incoming wave. There is a node one quarter of a wavelength from the reflecting coast, another at three-quarters of a wavelength, then five-quarters, and so on.

The sloping water surface in the standing waves drives accelerations in the water beneath. When it is high water at the coast A, the slope accelerates water offshore, towards B, and the flow (the ebb tide) reaches its maximum three lunar hours later when the water surface is flat. The water surface then starts to slope down from B towards A and the offshore flow is decelerated, becoming slack when it is low water at A. The currents and elevations in a standing wave are therefore out of phase. Places where the tide behaves as a standing wave (e.g. the northern Irish Sea in Europe) experience slack water at high and low tide with maximum current halfway between.

It is a characteristic of a standing wave that the extent of the up-and-down movement of the surface (the tidal range) varies from place to place, from zero at the node up to a maximum at the antinodes. High water occurs everywhere between two adjacent nodes at the same time, and when it is high tide on one side of a node it is low tide on the other side. The fastest currents occur at the nodes.

Resonance

Figure 20 allows us to anticipate the reason why we can get large tides in shelf seas. If the progressive tide wave travelling from the shelf edge to the coast is reflected and bounces back without any loss of energy, a standing wave tide will be created on the continental shelf. The movement of the standing wave at the edge of the shelf must match that of the ocean tide. If the edge of the shelf is close to a node in the standing wave, a small tide in the ocean will produce much larger tides at the antinodes. In particular, there will be a large tide at the coast, which is always located at an antinode of the standing wave. The shelf sea is said to be in resonance with the tidal forcing and there is amplification of the tide from the ocean to the coast.

The condition for resonance is that the width of the shelf should be an odd number of quarter wavelengths of the tide. For a shelf sea of depth 100 metres, the required width to be in quarter-wavelength resonance with the M_2 tide is 350 kilometres (a shelf twice as wide would be in resonance with the diurnal tide). Several of the world's shelf sea areas, particularly in South

America, north-west Europe, north-eastern America, and the Arctic are close to resonance with the semi-diurnal tide. These places have large tides and dissipate a disproportionately large quantity (compared to their size) of the world's tidal energy through friction.

What happens, exactly, when a shelf sea is in resonance? How can a small tide at the edge of the ocean produce a much larger tide hundreds of kilometres away? The rise and fall of the ocean at the edge of the shelf can be thought of as a paddle, pushing water onto the shelf when the ocean tide is high, and pulling water off the shelf when the ocean tide is low. In a resonant shelf sea, this motion is matched by the natural period at which water flows on and off the shelf. The ocean tide continually feeds energy into the shelf sea, and the tide on the shelf grows until its height is limited by friction.

Localized resonance can also occur in part of a shelf, for example a large bay or gulf. If the length of the bay from the head to the mouth is close to the resonant length, a relatively small tide at the mouth will produce a much larger tide at the head of the bay. A celebrated example of this kind of resonance occurs in the Bay of Fundy in Canada. In this bay, a mean tidal range of 5 metres at the mouth is amplified to nearly 12 metres at the head. The amplification factor is not so great—about 2.4—but, acting on what is already a big tide at the mouth, resonance produces exceptional tides in the upper reaches of the Bay of Fundy.

An example of a resonant shelf sea

The capital of South Australia, Adelaide, lies on the eastern shore of a shallow gulf, Gulf St Vincent, which is connected to the open sea by Investigator Strait and the smaller Backstairs Passage (Figure 21). The mean depth of water is about 30 metres and the speed of a progressive tide wave is 62 kilometres per hour. The wavelength of the M_2 component of the tide is 767 kilometres and one quarter of a wavelength is 192 kilometres. The distance from the head of the gulf to Stenhouse Bay at the western entrance of Investigator Strait is about 180 kilometres. The system is therefore close to quarter-wavelength resonance with the semi-diurnal tide and we would expect to see the tide amplified within the gulf.

Figure 21 shows co-tidal and co-range lines for the M_2 tide in the region. The amplitude increases from 17 centimetres at the western end of Investigator Strait to 62 centimetres at the head of Gulf St Vincent, an amplification factor of 3.6. Within Investigator Strait the tide has the appearance of a progressive wave: the high water moves from west to east, taking 1.5 hours to travel the

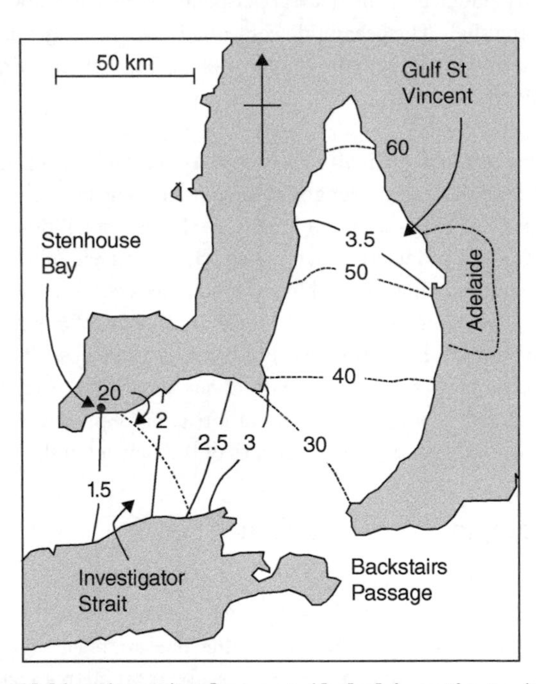

21. Co-tidal (continuous) and co-range (dashed, in centimetres) lines for the M_2 tide in the approaches to Adelaide, South Australia (based on Bowers and Lennon, 1990).

length of the strait. In Gulf St Vincent, the tide is a standing wave: high water occurs at the same time (within half an hour or so) over the whole area of the gulf, and the tidal range increases towards the head of the gulf. The tidal streams act in unison throughout the area, turning at the times of high and low tide at the head of the gulf.

The transition from a progressive wave at the entrance to a standing wave at the head of the bay is observed in many semi-enclosed seas. In between these regimes the tide is in transition from one kind of wave to another: it is part standing and part progressive. The behaviour can be understood by allowing for the effect of friction. A progressive wave loses energy through friction, first as it travels towards the shore, and then again after it is reflected and is travelling in the opposite direction. The energy in a water wave is proportional to the square of its height (the vertical distance from crest to trough), so a loss of energy means that the wave height becomes smaller as the wave travels along. The easiest way to allow for this is to let the wave lose a constant proportion of its height for every kilometre it travels across the ground.

The shape of the water surface in the bay at any time can be found by adding the incoming and reflected waves. We can start at any time and draw the shape of the wave approaching the coast. The wave will decrease in height with distance towards the shore because of the effect of friction. We can then draw the shape of the reflected wave (which will decrease in height with distance offshore) and add the two waves to get the shape of the water surface. We then step forward in time by an hour. The incoming wave will move one hour's worth, or 62 kilometres, towards the shore, and the reflected wave 62 kilometres in the opposite direction. We again add these two waves to get the shape of the surface at this time. The procedure is repeated every hour through a tidal cycle.

Figure 22 shows the result. To match the shape to observations, we have let the wave height decrease by 0.05 per cent per

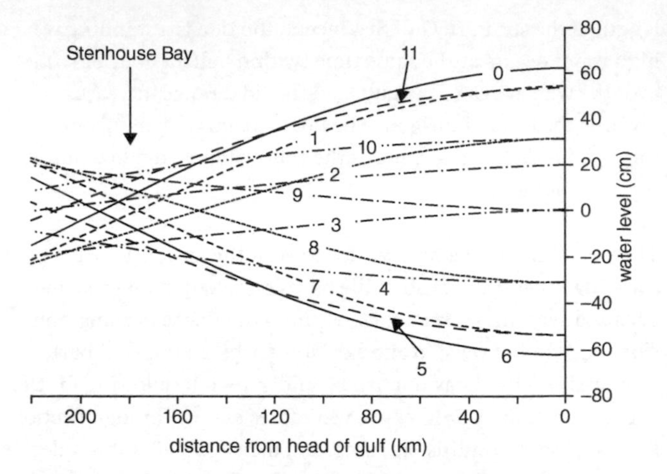

22. The shape of the water surface in Investigator Strait and Gulf St Vincent at hourly intervals (indicated by labels).

kilometre (the procedure teaches us something about the magnitude of tidal friction). The pattern shown in Figure 22 explains many of the features of the observed tide. There is an amplification of the tidal range from Stenhouse Bay to the head of the gulf. Within a distance of about 90 kilometres from the head of the gulf, high water occurs simultaneously—the tide is behaving as a standing wave. Further away from the head—in Investigator Strait—the time of high tide changes with distance, becoming earlier with increasing distance from the gulf head. Here, the tide is behaving as a partly progressive wave travelling inwards. High water at Stenhouse Bay occurs two hours before that at the head of the bay.

Near the head of the gulf, the incoming and reflected waves are of similar size and add together to produce a standing wave. Further offshore, the reflected wave is reduced to the point where it is considerably smaller than the incoming wave. Here, the incoming wave is dominant and this explains the progressive nature of the tide wave. Because the incoming and reflected waves are no longer of equal size at the node, they no longer cancel, and the node is replaced by a region of reduced tidal range, the extent of the reduction depending on the relative size of the two waves.

Most of the time, the water surface shown in Figure 22 is sloping. The slope creates a pressure gradient force which either drives accelerations or balances the effect of bed friction. At hour 0 the slope in Gulf St Vincent is creating an offshore acceleration which turns the flood into ebb (and vice-versa at hour 6). Hours 3 and 9 are the times of maximum ebb and flood current, and at these times the pressure gradient force created by the surface slope balances the effect of bed friction. The water surface slopes down towards the head of the gulf during the flood and in the opposite direction during the ebb.

Effect of Earth rotation

A detail of Figure 21 that is not explained by the account above is the fact that the 20 centimetres co-range line crosses Investigator Strait diagonally. The tidal range on the northern side of the strait is greater than that at a point opposite on the southern shore. This is a result of the Earth's rotation.

The effect of Earth rotation on a progressive tide wave is to create slopes in the water surface at right angles to the direction the wave is travelling. Imagine a progressive wave travelling down a channel in the southern hemisphere. At the crest of the wave, the current is flowing with the wave and the Coriolis effect pushes water towards the left hand shore of the channel, looking in the direction of wave travel. At the wave trough, the current is flowing against the direction of wave travel and the Coriolis effect pushes water towards the right hand shore. The result is a wave which has higher crests and lower troughs (and so a greater tidal range) on the left hand shore of the channel. In the northern hemisphere, the larger tidal range is found on the right hand shore looking down the direction of wave travel. This kind of wave is called a

Kelvin wave, after William Thomson, Lord Kelvin, born in Belfast in 1824 (the same Kelvin that we came across at the end of Chapter 3). It is this effect and the progressive nature of the wave in Investigator Strait that creates the larger tides on the northern shore.

When a Kelvin wave reflects off the head of a bay or gulf, the tide will behave as the sum of two Kelvin waves travelling in opposite directions. If the gulf is wide enough for Earth rotation effects to matter (that is, it takes a significant part of a tidal cycle for a wave to cross from one side to the other) an amphidromic system will form, with the crest of the wave (the high tide) sweeping around the gulf in an anti-clockwise direction in the northern hemisphere and clockwise in the south.

In the absence of friction, the amphidromic point will form at a point B one quarter wavelength from the head and in the middle of the bay, equidistant from the shores at C and D in Figure 23.

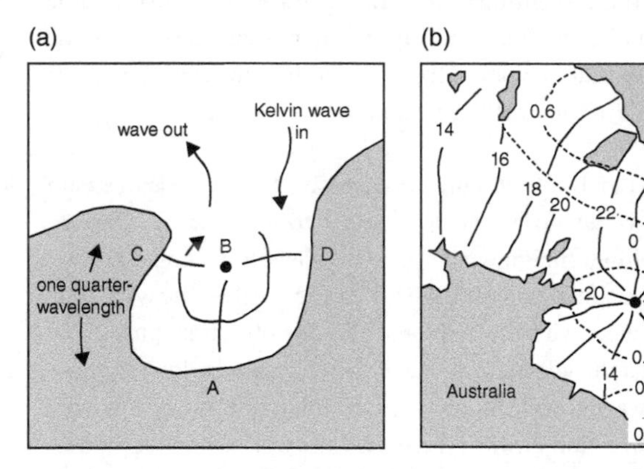

23. (a) A sketch of an amphidromic system, created by a Kelvin wave reflected off the head of a bay in the southern hemisphere (the amphidromic point is labelled B). (b) Co-tidal lines (continuous) for the diurnal tide in the Gulf of Carpentaria, marked with the time of high water in lunar hours and co-range lines (dashed) with tidal amplitude in metres (based on Webb, 2012).

The Gulf of Carpentaria, in northern Australia, is an example of a shallow sea which is close to quarter-wavelength resonance with the *diurnal* tide. The bay is about 500 kilometres wide and measures between 500 and 600 kilometres from its southern shore to where it joins the Arafura Sea and Torres Strait in the north. The mean depth of the bay is 60 metres and the speed of a progressive tide wave is 87 kilometres per hour. The length of the diurnal tide wave (with period twenty-four lunar hours) is 2,183 kilometres. One quarter of a wavelength is 546 kilometres and the amphidromic point of the diurnal tide is close to the mouth of the gulf. In contrast, the semi-diurnal tide (which has a wavelength half that of the diurnal tide) has an amphidrome located about 250 kilometres from the shore; there is no amplification of the semi-diurnal tide from the mouth to the head of the gulf.

A co-tidal chart of the diurnal tide in the Gulf of Carpentaria is shown in Figure 23(b). The high water travels around the bay in a clockwise direction about an amphidromic point located close to the mouth of the gulf. The tidal range is largest at the shore (particularly at the south-east corner) at points furthest from the amphidrome, and decreases to nothing at the amphidromic point. In the top left of this picture, the tide is behaving as a Kelvin wave. The high tide travels from west to east, and the tidal range is greatest on the northern shore (that is the left hand shore looking down the direction of wave travel).

Not all bays or gulfs exhibit an amphidromic system. First, the gulf needs to stretch, from the mouth to the head, a distance of at least one quarter of a tidal wavelength and, as we have seen, that can be a distance of several hundred kilometres. Only large bays will be big enough to contain an amphidrome. Second, the bay must be wide enough for Earth rotation to matter. In practice that means that the time taken for a tide wave to cross from one side of the bay to another must be a significant fraction of a tidal cycle. On a rotating Earth, a progressive wave will always be converted into a Kelvin wave, but the difference in water level from one side to the other will be small in a narrow channel.

The effect of friction and a poor reflection of the wave at the head of the bay will make the reflected wave smaller than the incoming wave. The two waves will no longer cancel along the centre line of the bay, and the amphidromic point moves to a place where the reflected wave is large enough to cancel the incoming wave. It moves to the left (in the northern hemisphere) and to the right (in the southern hemisphere) looking into the bay. In extreme cases, where the friction is very high, the reflected wave becomes too weak to cancel the incoming wave anywhere. The amphidromic point appears to move right out of the sea altogether, on to the adjacent land, and it is called a degenerate amphidrome.

Tides in shallow water

When a progressive wave enters water shallow compared to the wave height, the crest, being in deeper water, travels noticeably faster than the trough. As the wave travels up a shallow estuary (for example), the crest catches up with the preceding trough and the front face of the wave is steepened. As British physicist Lord Rayleigh put it:

> When waves advance over shallow water…the crests tend to gain upon the hollows, so that the anterior slopes become steeper and steeper and the posterior slopes more gradually sloping.

The same effect is seen in estuaries where the tide is a standing wave, but here the process is different. The tide floods into the estuary when it is high tide in the open sea, and the greater depth of water reduces the effect of bottom friction. The flood current is faster than the ebb and, since equal volumes of water (equal to the *tidal prism*) must enter and leave the estuary over a tidal cycle, the duration of the flood is shortened relative to the ebb. An example of the quick rise and slow fall of the tide in an estuary can be seen in Figure 12.

The distortion of the tidal curve in shallow water presents a problem for the accurate prediction of the times of high and low waters. There is no longer a fixed time interval between a high tide and the following low tide. This effect cannot be simulated using just the harmonics that are part of the tide-generating force. Harmonics with periods equal to a fraction of the semi-diurnal tidal period are needed to produce the shape of the tidal curve. Tidal analysts have a toolbox of such *higher harmonics* that they can fit to the observed tidal curve to create the required shape and produce accurate predictions of both high and low tides in shallow water.

A rare feature of the tide in shallow water, seen at just a few locations worldwide, is a double high water or its close relative, a double low water. In a double high water, the tide rises to a first high, falls for a short while, and then rises again to a second high before finally falling to a single low tide. The most famous example is found at Southampton on the south coast of England where the extended period of deep water around high tide was important to the development of the port. Figure 24 shows an example of a small Scottish port which has a double high water.

Double high waters can be made by adding higher harmonics to the semi-diurnal tidal curve. If the higher harmonic has a trough or low water at the time of the high in the M_2 tide, a double high water can be made if the harmonic has a large enough amplitude. The reason that double tides are so rare is that the higher harmonics are not usually big enough for the job and they don't have the right phase relative to the principal tidal constituent. Port Ellen lies near an amphidrome of the semi-diurnal tide and the shallow-water harmonics (particularly one with a period of four lunar hours) are large enough to create a double high water at neap tides. It is often the case that, on the other side of the

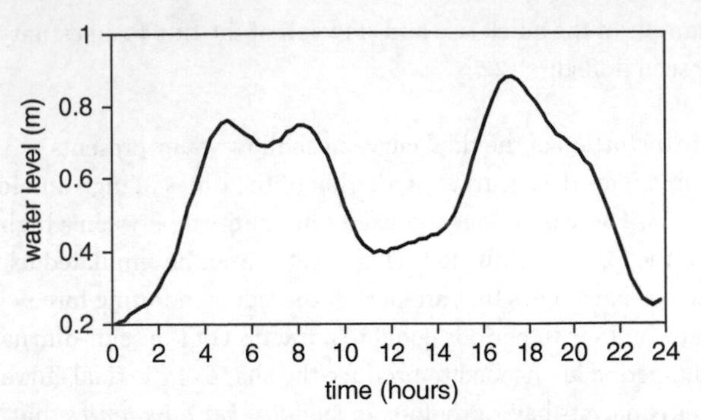

24. The double high water at Port Ellen in Scotland on the morning of 19th February 2010.

semi-diurnal amphidrome from the double high tide, there is a double low tide to be found. For example, on the eastern coast of the United States, a double high tide is observed at Woods Hole in Massachusetts and a double low tide at Providence, Rhode Island.

Chapter 5
Tidal bores

A *tidal bore* (Figure 25) is perhaps the most spectacular tidal phenomenon that can be readily observed. It is truly a 'tidal wave' (a term often used incorrectly to describe *tsunamis*, which are not tidal phenomena). When a large tide enters a shallow, funnel-shaped estuary with a gently sloping bottom, its waveform is distorted and this can lead to an impressive rolling 'wall of water' travelling upriver. The word bore is thought to be derived from the Old Norse *bára*, meaning 'wave' or 'billow'—apt, given their dynamics and appearance—and for tide enthusiasts (and many surfers) a well-developed tidal bore is the holy grail of natural occurrences.

The formation of a tidal bore

Tidal variations in water level at the mouth of an estuary create a wave that propagates upstream. This wave is gradually modified by changes in estuary width and depth, by friction with the bottom, and by the river flow seaward (Figure 26). Estuaries that narrow and shoal steadily with distance inland have a funnelling effect that amplifies the tidal wave, increasing its amplitude. In fact, the narrowing of the estuary is slightly more important to this amplifying effect than the shoaling.

25. A tidal bore (a) on the River Dee (UK) and secondary waves, or 'whelps' (b), behind the bore front.

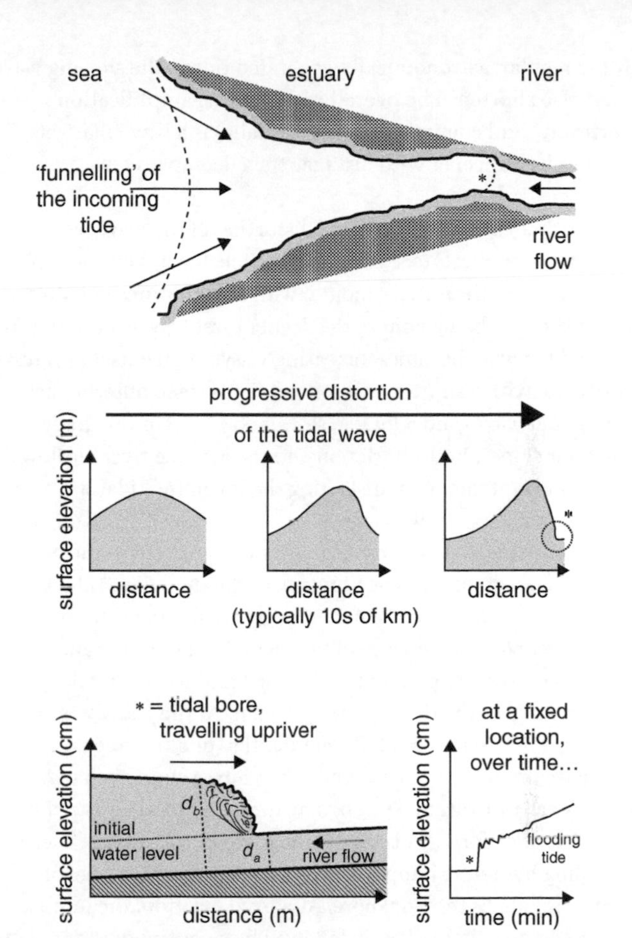

26. Funnelling and progressive distortion of an incoming tidal wave in an estuary. At locations where both the tide and the estuary geometry are suitable, a tidal bore may form (denoted by an asterisk) and travel upriver.

The tidal wave slows as it moves into progressively shallower water. Since there is a fixed relationship between the speed of a wave, its wavelength, and its frequency (speed is wavelength multiplied by frequency) and the frequency is constant (being set by the regular, astronomically-generated tides), the slowing wave must also shorten. The overall result of this amplification (vertically) and compression (horizontally) is a wave that gets progressively steeper with distance travelled upstream.

One further consideration is the distortion of the wave by shallow-water effects (see Chapter 4). While the waveform is initially symmetrical at the estuary mouth (with the durations of both flood and ebb stages being equal), the deeper crest travels particularly fast and the wave becomes increasingly asymmetric as it progresses upstream, with a shorter, faster flood (i.e. a steep anterior slope to the tidal wave) and a longer, slower ebb (i.e. a more gradual posterior slope). Bottom friction and the adverse river outflow are also important factors in distorting the incoming tidal wave.

Where the tide and estuary geometry conspire to produce a strong flood-dominated asymmetry in the incoming tidal wave the stage is set for a tidal bore. For a gently sloping estuary bed, the steepened water surface of the incoming flood tide and its associated pressure gradient drive a fast and somewhat deeper flow (relative to the river level) upstream. At the transition between the two flows of different depths (one riverine and downstream, the other tidal and upstream) a sudden change in water level, or hydraulic jump, may form and travel upstream as a distinct waveform. This 'hydraulic jump in translation' (i.e. a travelling hydraulic jump), which marks the leading edge of the incoming tide, is the tidal bore. At a fixed location, the jump is observed and is immediately followed by a continuous (and usually rapid) rise in water level as the flood tide progresses.

Where and when to see a bore

Estuary shape is clearly important to the formation of a tidal bore, and a large tidal range is also crucial. The precise range required varies from place to place, but few bores are known to form where the tidal range is less than about 6 metres. For this reason, the geographical distribution of tidal bores closely matches that of large tides, particularly those belonging to a semi-diurnal tidal regime (diurnal tides are rarely steep enough to form bores—they have a longer wavelength, typically smaller amplitudes, and the initial rise in water level is slow enough that the water surface responds 'normally'). The coasts of north-west Europe (particularly France and the United Kingdom), the Bay of Fundy in Canada, the north-east coast of South America, parts of the north coast of Australia, the Gulf of Alaska, and the East China Sea are tidal bore hotspots.

Around half of known tidal bores occur in rivers adjacent to basins that resonate with the tide. That isn't to say such rivers host tidal bores with every tidal cycle. In many cases the very largest tides are needed for a bore to develop. Spring tides and particularly those occurring around the equinoxes are generally good times to observe tidal bores.

The timing of a bore can be predicted with some certainty, and its anticipated time of arrival is usually given relative to the predicted time of high water at a nearby port. As we have seen, the high waters of spring tides always occur at around the same times of day for a given location, and so bores also occur at more or less fixed times for a given site. A word of warning is needed though: the speed of the bore (and its remarkability) can be affected by wind conditions, barometric pressure, and rainfall (which influences river flow/depth). The River Dee Bore in the United Kingdom, for example, frequently arrives at key observation points up to thirty minutes early or late. For the tidal bore observer or surfer, arriving earlier than the predicted time is best to avoid disappointment.

Types of tidal bore

Not all bores are equal. They come in a range of different strengths and sizes from place to place (from just a few centimetres to 6 metres in height). Even at the same site, they vary from one bore to the next. They also come in different forms, from smooth, non-breaking 'undular' waves (generally the most common and often experienced first as an apparent 'buckling' of the river surface), through a variety of breaking forms of increasing violence, and culminating in those with very odd behaviours (such as the generation of large jets of water extending ahead of the bore front).

The bore front represents a travelling discontinuity of water depth and flow velocity, which has historically elicited much interest among hydraulic engineers and applied mathematicians. The English engineer, hydrodynamicist, and naval architect William Froude (1810–79) is credited with a useful parameter for describing the likely character of a tidal bore. The tidal bore Froude number (*Fr*) is defined in a coordinate system that moves with the bore front (effectively bringing it 'to rest' for the sake of simplifying the mathematics) and is dimensionless (i.e. it has no physical units). It is the ratio of the bore speed (the bore's observed upstream speed summed with the adverse downstream river speed) to the natural speed of a hypothetical 'shallow-water' *gravity wave* (see Glossary) travelling freely on the undisturbed river (i.e., a speed $\sqrt{(gd_a)}$, where g is acceleration due to gravity and d_a is the water depth of the undisturbed river, ahead of the bore). For $Fr < 1$, the flow is said to be *subcritical* and no bore develops. For $Fr > 1$, the flow is *supercritical*, and various types of bore can form (see Table 3). As the ratio suggests, a tidal bore represents a hydrodynamic shock to the river system, with a wave being forced to travel faster than it would naturally in that environment. The Froude number can also be shown to depend on the ratio of the water depths behind and ahead of the bore (d_b/d_a; Figure 26): the greater this ratio, the greater the Froude number, and the stronger and more spectacular the bore.

The dependence of the Froude number on the ratio of depth behind to that ahead of the bore (d_b/d_a) tells us something about

Table 3. Types of tidal bore

Froude number, *Fr*	Breaking/ non-breaking	Bore type	Description
1.0–1.7	Non-breaking	Undular	Smooth primary wave followed by a train of secondary waves, or 'whelps'.
1.7–2.5	Breaking	Weak jump	Bore front breaks, but not violently.
2.5–4.5	Breaking	Oscillating jump	Violent front, penetrated by upstream flow as oscillating jets; large surface waves generated behind the front.
4.5–9.0	Breaking	Steady jump	Well-formed breaking bore with effective encrgy dissipation and no additional surface waves formed.
> 9.0	Breaking	Strong jump	Large, high-speed jets of water extend ahead of the turbulent front.

how the bore front might vary *across* the river. Near the banks, where the river is generally shallower, this ratio is greater, explaining why bores are frequently observed to be breaking near the banks but undular in the mid-channel (where the ratio is smaller). It also explains why bores may change their form as they travel upstream, for example from undular to breaking and back, responding to variations in river depth.

River beds with typical bank-to-bank cross-sections often cause bores to develop a curved shape, viewed from above. The near-bank parts of the bore propagate more slowly than the centre because they are in shallower water and the wave undergoes *refraction* (at both ends). At the outer edges of the river channel, the bore gains components of velocity directed towards the banks, approaching them obliquely with the potential for additional breaking to occur.

Whelps

Non-breaking, or undular, tidal bores often have a train of waves of decreasing amplitude following the lead wave (Figure 25(b)). In the UK, these secondary waves are referred to as *whelps*, whereas on the River Seine (France) they were known as *éteules*, or 'stubble' (possibly a reference to the appearance of a ploughed field after the harvest). Since they are of lesser height, these waves tend to move more slowly than the lead wave and are gradually left behind to eventually dissipate (physicists refer to such a wave packet as being weakly dispersive). Whelps can be seen for some time after the bore front's passage, in some cases more than thirty minutes. They appear increasingly chaotic with time, partly because whelps are also refracted by the shoaling of the channel towards its banks and reflect off the banks to interact with each other behind the progressing bore.

The rumble

Large tidal bores produce noises that can be placed on a spectrum between 'menacing rumble' and 'deafening roar'. In fact, the noise results from a cacophony of sounds produced by various physical processes: *turbulence* and *entrained* air bubbles in the bore front (where the bore breaks); the movement of sediments at the river bed; and interactions with obstacles to the flow and with the banks.

The first scientific field measurements of the rumble sound were carried out at the turn of the century in France by tidal bore expert

Hubert Chanson. Chanson's work identified three phases with distinct acoustic characteristics: first the approaching bore, with a sound intensity increasing with time; then the passing bore front, with a noise on average five times louder than the previous phase; and finally the upstream propagation of the bore, with the sounds of the flood tide behind it. Analysis showed most of the sound's energy to be concentrated around the frequencies of 74 hertz and 131 hertz, perfectly audible to humans (the frequency range of the human ear is approximately 20 hertz to 20 kilohertz) but rather low in pitch. As Chanson points out, this is lower in pitch than the beating of a bass drum or the labouring of a locomotive train. The acoustic spectrum also showed a secondary peak at higher frequencies (8–10 kilohertz). Current thinking is that the low pitch rumbling is caused by air bubbles trapped in the turbulent bore front, while the weaker, higher pitch component may be produced by sediments being moved along the river bed (and in particular by particle collisions) beneath the bore.

The low pitch component of the tidal bore rumble is its defining characteristic. It is also the reason why so many bore anecdotes focus on how far away the bore could be heard, or how far in advance of the bore front's arrival it could be anticipated. Low frequency sounds tend to travel further than higher frequency ones: less energy is 'lost' to the medium they travel through. Classic examples are the use of low frequency sound for fog horns and the low pitch thunder that may be heard emanating from distant lightning storms. All this adds up to stories of impressive distances covered by tidal bore rumbles.

Not all stories focus on distance, however: some concern direction. Waiting for one of the Bay of Fundy's tidal bores in the fog for which the region is famous is reported to be an unnerving experience. Under these conditions, the rumble of the bore can be heard but, rather eerily, the direction of approach cannot easily be determined.

Death of a tidal bore

Like all good things, a tidal bore must end. As a bore propagates, its kinetic energy is lost owing to friction with the river bed, and to the viscosity of the water itself. The loss is augmented by any shore-break occurring as a result of refraction onto the river banks. Some energy is used in producing the rumble sound and some in moving sediments on the river bed. Where whelps are present, there may be some backward radiation of the bore's energy (in the downstream direction relative to the progressing front). Generally, the energy passes to a cascade of turbulence on finer and finer scales, and ultimately the bore's energy is dissipated as heat.

The energy loss is sensitive to the difference in water depth behind and in front of the bore. In fact, it increases in proportion to this difference cubed so that a large tidal bore tends to lose energy much faster than a smaller one. As a consequence of this energy loss, the bore slows down and at the point at which the bore speed equals the river's flow speed it is halted. Any remaining waveform can even be carried back downstream by the river.

With access to a car and traffic-free roads, the death of a bore can be seen in action. We have ourselves witnessed impressive tidal bores in the UK at Saltney Ferry Bridge on the River Dee and at Minsterworth on the River Severn, only to be disappointed a short time later further inland near Chester Weir and Maisemore, respectively.

Hall of fame

In 1988, Susan Bartsch-Winkler and David Lynch from the United States Geological Survey (USGS) produced a catalogue of known tidal bores. It incorporated sixty-seven bores, in sixteen countries, spread across every continent except Antarctica. Many more are predicted to exist where conditions are suitable, but are either undiscovered or unreported. Humans worldwide have assigned fantastically evocative names to their local bores (Table 4) and they pervade cultural rituals and historical accounts. 'Eagre' (or 'Aegir') is a name used traditionally in the United Kingdom for bores on the rivers Trent, Ouse, and Severn. It has an uncertain etymology, but our favourite (and the most romantic) theory is that it is derived from the name of a *jötunn* in Norse mythology (i.e. the giant *Ægir*, king of the sea), given early Scandinavian influence on the British Isles. Bores even appear in classical literature (e.g. George Eliot's *The Mill on the Floss* and Jules Verne's *Eight Hundred Leagues on the Amazon*).

One of the most impressive tidal bores occurs on the Qiantang River in China. Named the 'Silver Dragon', the Qiantang Bore can attain a height of 4 metres and speeds of up to 12 metres per second. The bore travels around 100 kilometres upriver from the funnel-shaped estuary in which it forms. Captain William Usborne Moore, a British Admiralty hydrographer whose efforts to survey the area in the late 1800s were severely hampered by the bore, noted that its rumble could be heard an hour ahead of its arrival.

The Qiantang bore is famously associated with a myth that it was unleashed as a punishment upon a 5th-century BC emperor who had ordered the suicide of a popular general, Wu Tzu-Hsü, and had his body thrown into the river. However, by the 1st century AD Chinese scholars had already noted the correlation between the bore's behaviour and the phases of the Moon. By 1056 AD, a table predicting the occurrence of the bore had been constructed, making it arguably the first ever tide table. The Silver Dragon is something of a spectacle, which today attracts tens of thousands of visitors a year. There is even an annual Qiantang Tidal Bore Watching Festival, which takes place around the autumnal equinox. The bore still manages to surprise, however: it regularly overtops levees, sweeping away onlookers and causing fatalities.

Table 4. Some well-known tidal bores

Country	River	Tidal body	Common/local name	Height; speed
Australia	Daly	Timor Sea	–	1.5 m; 4–5 m/s
	Styx	Broad Sound	–	–
Brazil	Amazon	Atlantic Ocean	*Pororoca*	1–3 m; –
	Araguari	Atlantic Ocean		2 m; –
Canada	Petitcodiac	Bay of Fundy	–	1–1.5 m; –
	Shubenacadie	Bay of Fundy	–	0.3 m; –
China	Qiantang	East China Sea	Silver Dragon	1–4 m; 5–12 m/s
France	Dordogne	Bay of Biscay	–	–
	Garonne	Bay of Biscay	–	–
	Seine	English Channel	*La Barre*; *Le Mascaret*	Reduced (see text)
India	Hooghly	Bay of Bengal	*Bán*	2 m; 7.5 m/s
Indonesia	Kampar	South China Sea	*Bono*; Seven Ghosts	–
Malaysia	Batang Lupar	South China Sea	*Benak*	1 m; 5 m/s
Mexico	Colorado	Gulf of California	*El Burro*	Reduced (see text)
Mozambique	Pungwe	Indian Ocean	–	0.7 m; –

Pakistan	Indus	Arabian Sea	–	2 m; –
Papua New Guinea	Fly	Coral Sea	*Ibua*	2 m; –
United Kingdom	Dee	Irish Sea	–	0.2–0.5 m; 4 m/s
	Mersey	Irish Sea	–	–
	Severn	Bristol Channel	–	1–2 m; 2–6 m/s
	Trent	North Sea	*Aegir*; *Eagre*	1–2 m; 6 m/s
USA (Alaska)	Turnagain Arm	Cook Inlet	–	1.5 m; >4 m/s

In Brazil, the Amazon River and its tributaries host tidal bores known as *Pororoca*. The name is thought to be derived from a word meaning 'great roar' in the language of the indigenous Tupi people. Albert Defant, a physical oceanographer, claimed the sound of the main Amazon bore could be heard at a distance of 22 kilometres. During its passage it renders the river impassable, but it is the sheer scale of the bore that makes it particularly noteworthy. The bore front is, at points, kilometres wide and owing to the Amazon's gently sloping bed it penetrates further inland than any other bore, up to 800 kilometres. An impressive train of whelps follows the lead wave, with several observers reporting many tens of large whelps disappearing over a distant horizon.

Some bores on the Amazon's tributaries are unusual in that they effectively form more than 100 kilometres inland. By contrast, Brazil's Araguari River to the north hosts a bore that forms up to 10 kilometres *offshore*, aided by the shoaling of the incoming tide over extensive delta deposits.

The River Severn in the United Kingdom has a tidal bore that has become a famous surfing attraction. At the landward end of the Bristol Channel (which has one of the largest tidal ranges in the world), the Severn Bore was first surfed in 1955 by Lieutenant-Colonel 'Mad Jack' Churchill (whose nickname was earned rather more for his military exploits than for his daredevil surfing). It has been surfed ever since and has often been the bore ridden in Guinness World Record attempts for 'longest surfing ride on a river bore' (distances surfed standing up on the bore have been over 12 kilometres, taking well over an hour). The bore occurs on about 130 days per year (usually twice per day—morning and night), occasionally achieving heights greater than 2 metres and speeds of around 6 metres/second. Severn Bore predictions are assigned star-ratings, with '5*' representing an exceptional bore and attracting large crowds. It's good to know it is not just oceanographers that get a little over-enthusiastic about tidal bores.

Tidal bores and people

Tidal bores can be useful. Their strong flood currents represent the fastest way to pass upriver. They are surfed for leisure, and are woven into the fabric of diverse cultures worldwide. On the other hand, they can be a nuisance to shipping and at worst just plain dangerous. The River Seine (France) formerly hosted a substantial bore (Figure 27), or *mascaret* in French, known as *La Barre* or *Le Mascaret*, which had a sinister reputation for wrecking ships and causing human fatalities. Alexander the Great, 4th-century BC leader of the Greek kingdom of Macedon, is thought to have fallen foul of a tidal bore on the Indus River in modern-day Pakistan. Alexander's fleet was stranded by a violent wave surging upriver, only to be re-floated about twelve hours later by a similar but smaller incoming tide.

27. Bore on the River Seine before its near elimination in the 1960s.

Human influence on tidal bores to date has been largely destructive. The Seine Bore has been almost eliminated by waterway engineering and dredging that has altered the river mouth. A once powerful bore on the Colorado River in Mexico (known as *El Burro*, or 'the donkey') is now much reduced by patterns of dredging and silting at the mouth. The bore on the Petitcodiac River (Bay of Fundy, Canada) all but disappeared after the construction of a causeway. The River Severn Bore will likely be subject to the same fate if certain proposed schemes to harness tidal energy were to go ahead.

Projected sea-level rise is likely to alter the picture somewhat with respect to the distribution of tidal bores. It is difficult to say how exactly, but coastal tidal regimes and estuarine morphologies will change with sea-level rise and some known bores may disappear while new bores may emerge elsewhere.

Impact on estuarine processes and ecosystems

Tidal bores influence physical processes in their host estuaries, and in turn affect the ecosystems they support. They present unique flow conditions and are associated with turbulent mixing and the transport of suspended material upstream, both as larger objects (e.g. logs, stunned fish) and as clouds of turbid, sediment-laden water. They scour the river bed, periodically redistributing sediments, and sharp jumps can also be seen in water temperature and salinity following the passage of a bore.

The ecological importance of tidal bores can be seen in the behaviours they elicit in coastal and marine animals. Diverse organisms have been reported (scientifically and/or anecdotally) to feed immediately behind a tidal bore: piranhas (Amazon River, Brazil), bald eagles and beluga whales (Turnagain Arm, Alaska), sharks (Broad Sound, Australia), and crocodiles (Daly River, Australia) are all believed to use tidal bores to their advantage. Hubert Chanson reports that swans regularly ride upstream on the bores of the Garonne and Dordogne Rivers (France), though whether this is out of pleasure or convenience it is difficult to say.

Bores in other settings

Bores can form under circumstances different to those described above. A bore may form in a river that does not typically host one if it is affected by a tsunami or extreme weather event. A bore-like wave can even form on broad sand flats where there is a very large tidal range. The Baie du Mont St Michel (France) hosts such a feature, which is famed for travelling faster than a horse can gallop. 'Internal bores' are known to form when *internal tides* (i.e. tidal perturbations of the interface between layers of different density in the ocean—see Chapter 6) break on the outer margins of continental shelves.

Bore-like features have also been reported for various layers in the atmosphere. For example, undular bores in the troposphere are believed to be responsible for the 'morning glory' clouds seen over northern Australia's Gulf of Carpentaria, and atmospheric bores have been measured near thunderstorms on the central plains of the USA. Such atmospheric undular bores may even be occurring on Mars, and have been proposed as an explanation for long, linear clouds sometimes seen along the flanks of the Tharsis volcanoes.

Waves generated by the rush of water from a burst dam ('dam-break waves') also take the form of bores. In fact, such a wave has a key role in D. H. Lawrence's short novel *The Gypsy and the Virgin*. In Christopher Miles's film adaptation of the story the wave is actually 'played' by the Severn Tidal Bore, which makes a cameo appearance presumably because it was less expensive and more predictable than the bursting of a dam.

Tidal races

An impressive hydraulic phenomenon, known as a tidal race, is an area of rough water created by fast tidal streams flowing over a shallow, uneven seabed. Significant races are marked on charts as navigational hazards. Some of the more accessible ones attract kayakers who enjoy the steep, fixed water surface with currents flowing up the face of the wave.

When a flow passes over an underwater precipice it is reasonable to suppose that the flow will slow down in the deeper water beyond the precipice (Figure 28).

In order to decelerate the water, there must be a force acting to slow it down. As is usual with tidal currents, this force is provided by a sloping water surface which creates a pressure gradient force. The water surface rises to create the slope as the water flows over the precipice, making a wave that is fixed in position.

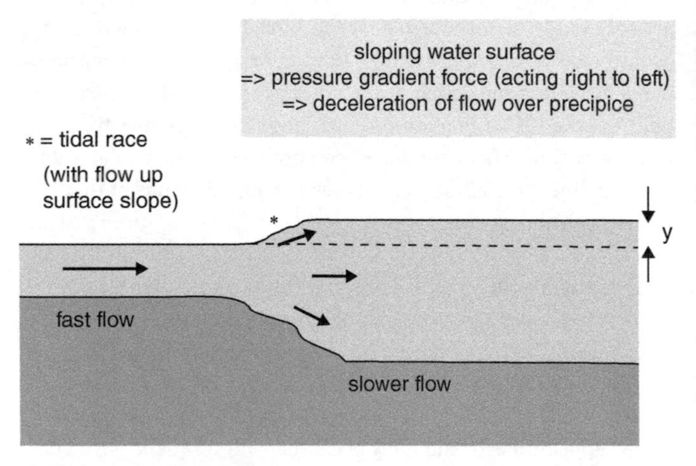

28. Formation of a tidal race.

At the jump, the water is flowing 'uphill'. The attraction for a kayaker is that they can hold themselves on the face of the wave with their weight balanced by the drag of the water carrying them upwards. The stationary wave, once formed, can last for an hour or more. Famous examples occur at The Bitches in west Wales (UK) and Race Rocks in British Columbia (Canada). When the tide turns so that the water is flowing up the underwater precipice, the surface steps down into the shallower water (so the surface keeps the same shape as in the illustration). Now, however, the flow is going 'downhill' (i.e. down the surface slope) and so is no good for holding a kayak in position.

Chapter 6
Tides and the Earth

The energy in tides comes, ultimately, from the spin of the Earth. This may surprise you, but it is the Earth turning within the pattern of lunar and solar tidal forces that drives the daily and semi-daily tidal motion. Tidal streams, rubbing against the seabed, lose energy through friction and, to make up for this loss, energy is transferred into the tide from the Earth's spin. As a result, our planet's rotation is gradually slowing and the day is lengthening. Most tidal friction happens in shelf seas, where the currents are strongest and the water is shallow, but there is an additional loss of energy in the body of the deep ocean, through the creation of waves called internal tides.

The slowing planet

Evidence for the slowing of the Earth's rotation comes from a number of sources, including the fossil record. Fine microscopic analysis of growth bands in fossilized marine corals shows that 400 million years ago there were 400 days in a year. If the length of the year was the same then as now, the daylength at that time must have been 21.9 hours. The length of an Earth day has therefore increased by 2.1 hours in 400 million years. This is equivalent to adding 1.9 milliseconds to the length of the day each century.

A further source of evidence for increasing daylength is the historical record of solar eclipses. Eclipses were (and are) important events and have been recorded by several ancient civilizations. The oldest known recording of a total solar eclipse took place in Babylon in 720 BC. As we go back through historical time, the speed at which the Earth was spinning increases enough to have a measurable effect on the position of places relative to the fixed stars.

To see how this works, imagine two planet Earths, one of which spins at a constant speed and the other with a speed which increases as we move back in time. A place on the Earth which speeds up as we turn back the clock will move ahead (that is eastwards) of its equivalent position on the constant-speed Earth. If we know that, at a particular time, a total solar eclipse was observed at (say) Babylon, it is possible to fix the position of Babylon, relative to the Sun and Moon, at that time. In the case of the 720 BC eclipse, we know that the accelerating Earth must have made a further quarter-turn, compared to its steady-speed counterpart, for Babylon to catch the eclipse. A quarter-turn in 2,700 years is equivalent to an increase of daylength of 1.8 milliseconds per century.

An accurate way of measuring the present deceleration of the Earth's spin is through the change in the distance to the Moon. As the Earth's spin slows, the conservation of angular momentum of the Earth–Moon system requires that the Earth's angular momentum is transferred to the Moon. The mean distance to the Moon and the length of the month are both gradually increasing. Laser ranging (using reflectors left at the time of the Moon landings) gives an accurate measurement of the distance to the Moon: the distance is increasing at a rate of just under 4 centimetres per year. The rate of loss of Earth's angular momentum that this implies equates to an increase in daylength at the present time of 2.4 milliseconds per century.

The observations of the Moon's increasing orbital radius tell us that the Earth is losing kinetic energy at a rate of 3.7×10^{12} watts, or 3.7 terawatts. If this energy is used to replace frictional losses by the tide, this figure should match what observations at sea tell us about tidal friction.

Tidal friction in shelf seas

The obvious place to look for the effects of tidal friction is in the resonant shelf seas, where the currents are fastest and friction is greatest. The first person to do this correctly was the British physicist Sir Geoffrey Ingram Taylor. Taylor realized that any analysis of tidal friction would have to allow for the fact that tidal currents are *turbulent*, and the turbulence will increase the effect of seabed friction on the flow.

In a tidal current in a shallow shelf sea, the water directly in contact with the sea floor is slowed down by friction. This water then slows down the water immediately above it and so on towards the surface, creating a profile of velocity increasing with height (Figure 29). A flow in which layers of water slide over each other at different speeds in this way (the flow is said to be *sheared*) is inherently unstable in water depths greater than a few centimetres. Some of the energy in the flow is used to create turbulence.

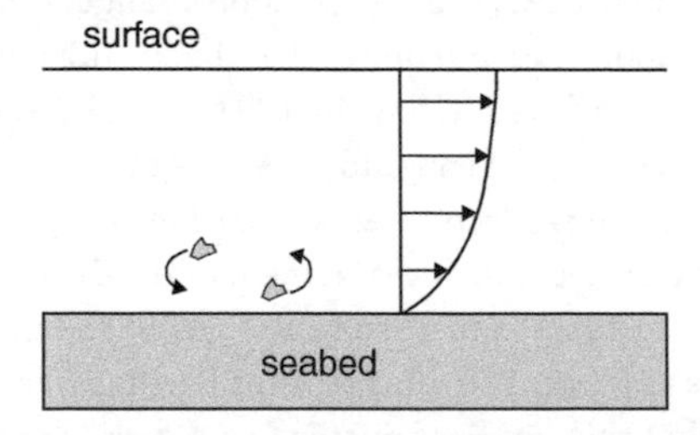

29. Turbulent exchange of parcels of water between layers in a tidal flow transfers momentum, increasing the height at which bed friction is felt and creating a velocity profile.

The turbulent eddies created by the tide move, in general, in three dimensions. The vertical eddies move water between the seabed and the surface. A way to picture this is that, each second, a parcel of water is transferred from one horizontal layer into an adjacent layer and, at the same time, an equal sized parcel is transferred in the opposite direction (we have sketched this exchange in Figure 29). There is no change in the volume of water in each layer but there is a sharing of momentum (and other water properties). The parcels of water from a faster moving layer speed up a slower layer and vice versa.

In the case of a tidal stream flowing over the seabed, the vertical exchange of horizontal momentum by turbulent eddies increases the height to which the friction of the seabed is felt. In a shallow sea with fast tidal streams it is not unusual to see an apparent 'boiling' motion at the surface caused by friction at the seabed.

In turbulent flows (in the atmosphere and the sea) the frictional force between the flow and a solid object depends on the square of the flow speed (a rule known as the quadratic friction law). The rate of energy loss from the flow equals the frictional force multiplied by the flow speed and so depends on the *cube* of the speed. A current flowing at 1 metre per second will lose energy one million times faster than one flowing at 1 centimetre per second. For this reason we would expect most energy losses due to bottom friction to occur in the resonant shelf seas (where speeds are of order 1 metre per second) rather than the deep ocean (where the speeds are of order 1 centimetre per second).

G. I. Taylor calculated the loss of tidal energy using the quadratic friction law with observed current speeds. His estimate of the energy lost by the tide in the Irish Sea, using a mean current speed of 1.17 metres per second, was $5x10^{10}$ watts, about 1.3 per cent of the total global energy dissipation rate of 3.7 terawatts predicted from astronomy.

Taylor checked his figure by calculating the flux of tidal energy flowing into the Irish Sea from outside. A progressive tidal wave carries energy; the flux of energy depends on the height and speed of the wave. Taylor realized that it was possible to calculate the net flux of tidal energy entering a semi-enclosed water body using measurements of currents and tidal elevations at the entrance. He used tidal diamonds to provide information about currents and shore-based tide gauges for the elevations. His estimate of the flux of tidal energy into the Irish Sea agreed (within the limits of experimental error) with his calculations of energy loss from bottom friction.

In the decades that followed Taylor's work, tidal frictional losses were calculated for a growing catalogue of shelf seas. As expected, the main areas in which energy is lost are those shelf seas close to resonance with the tidal forcing. Tidal friction is greatest, per unit area of sea, in Hudson Bay and the Labrador Sea of Canada, the north-west European Shelf, the Yellow Sea in China, the North Australian Shelf, and the Patagonian Shelf of South America. In total, for all shelf sea areas, the current estimate of the rate of energy loss through tidal friction is 2.5 terawatts.

This figure falls short of the 3.7 terawatts estimated from astronomy. If we are to match tidal friction to energy loss from the spinning Earth there must be a further drain on tidal energy which does not rely directly on bottom friction.

Internal tides

Oceanographers measure the structure of the ocean with an instrument called a CTD, which stands for conductivity-temperature-depth probe. The CTD is lowered using the ship's winch from the surface to as near the seabed as the operator cares, or dares, to go. To allow the sensors time to adjust to the environment through which they are descending, the CTD is lowered slowly: half a metre per second is the recommended speed. At a deep ocean station it can take well over an hour to profile from the surface to the bottom. On a long research cruise, the CTD soon becomes part of the routine work pattern, but for the first few profiles, scientists will gather round the computer screen to watch the plots of temperature, salinity, and other water properties unfold.

In a typical CTD profile in the deep ocean, there is a surface layer of uniform temperature, 10–100 metres deep, in which the Sun's heat is uniformly mixed by wind and waves. Below this, water temperature falls with increasing depth, rapidly at first and then more slowly. The steeper vertical gradient of temperature in the ocean marking the transition between warm water on top and cold water below is called the *thermocline*. As the ocean floor is approached, temperatures drop close to 0 degrees Celsius even in the tropics. Because warm water is less dense than cold water, the ocean thermocline is generally stable: the water density increases downwards. Occasionally, changes in salinity will contribute to the stratification (fresh water is less dense than salty), but in a stable state, density always increases from the surface down.

This density structure can be disturbed. If a parcel of water is displaced vertically in the thermocline it will experience a restoring force as it moves into water less or more dense than itself. The water parcel will oscillate about its equilibrium depth, making waves in the interior of the ocean known as internal waves. Internal waves may be generated by changes in the discharge of a major river into the sea, and by variable wind and atmospheric pressure effects on the surface of the ocean. They can also be made by tidal currents flowing over mountains on the sea floor. Internal waves with semi-diurnal or diurnal tidal periods are called internal tides. In the simplified case of an ocean with two layers of different density, these internal tides travel along the interface between the layers. If the vertical density variation is continuous over depth, internal waves can form at all depths and travel in directions at an angle to the horizontal.

It is fairly easy to make a small-scale internal wave in a drinking glass with salt water at the bottom and fresh water on top. You can stir some milk into either layer to make the density difference visible. The tricky part is making the layers without mixing them too much: it's best to put the light, surface water in first and then gently pour the denser water into the bottom through a drinking straw or small funnel. If you now tilt the glass and then set it straight, simulating the flow of the tide over an underwater hill, you will create an internal wave on the interface between the layers.

Internal tides travel slowly compared to surface tidal waves. This is because the difference in density, and hence the restoring force, between the layers in the stratified water is small compared to the difference in density between water and air at the ocean surface. Internal tidal waves travel at speeds of order 1 metre per second. The vertical movement in an internal tide wave can be large, however. The amplitude of internal tides in the ocean is typically tens of metres: several orders of magnitude greater than that of the surface tide. In a two-layer ocean, these vertical motions are largest at the density interface and decrease rapidly in amplitude with vertical distance from it. Near the sea surface, oscillations associated with a passing internal tidal wave are almost (but not quite) undetectable.

The earliest measurements of internal tides that we are aware of were made in the Kattegat by the Swedish oceanographer Otto Pettersson. In the Kattegat, salty water from the North Sea lies underneath fresher water from the Baltic, creating a sharp density interface. Pettersson fashioned a sinker which was dense enough to fall through the lighter upper layer but not dense enough to penetrate the heavier deep layer. The sinker therefore sat on the interface between the layers. Internal tidal waves then made the sinker move up and down, and the movements could be observed by fixing a light pole, long enough to stick out of the surface, to the sinker. The amplitude of the internal tide in the Kattegat was as great as 5 metres, much larger than the surface tide in these waters.

The horizontal currents generated by an internal tide produce alternating zones of convergent and divergent flow. These make patches of surface roughness which can be seen by eye. In some cases, particularly in coastal areas, natural and anthropogenic surface oils are swept together in the convergence zones to form 'slicks' that damp short surface waves (i.e. those which have surface tension as their restoring force—oceanographers call these *capillary waves*). More commonly, the interaction of capillary waves and the horizontal currents associated with internal tides concentrates these short waves in the convergence zones, locally *increasing* surface roughness (Figure 30(a)). In both cases, under light wind conditions the result is similar: clear, alternating bands of rough and smooth water.

The surface expressions of internal tides are visible in satellite radar images and astronaut photography (Figure 30(b)) as alternating bright and dark bands. Internal tides are observed to be generated in packets each tidal cycle and they travel large distances across oceans. Today, there are a number of innovative ways of studying internal tides, some of which have led to important scientific discoveries.

By combining satellite altimeter measurements of tidal amplitudes with model results for tidal currents, American oceanographers Gary Egbert and Richard Ray have calculated and mapped the global distribution of the rate at which energy from the surface tide is transferred to internal tides. Their method is an extension of that used by G. I. Taylor in the Irish Sea a century ago. The input of energy from the tide-generating force working on the current in a section of ocean is calculated. The net rate at which energy enters or leaves that section is determined from the elevations and currents on the boundary. The difference is the loss of energy from the surface tide.

30. Internal tides seen in sea surface roughness (a) and internal tide wave packets in astronaut photography (b).

The geographical distribution of the dissipation of surface tidal energy in the ocean leaves little doubt that it is being converted into internal tides. Most energy is dissipated at mid-ocean ridges and the continental shelf edge, where there are sudden changes in the depth of the ocean and where we would expect internal tides to be generated. The rate at which energy is put into the creation of internal tides is estimated to be at least 1 terawatt. The internal tide ultimately loses that energy by breaking at the edge of the ocean and by mixing the interior of the ocean.

And so we can balance the tidal energy budget of our planet. The Earth's spin is losing energy at a rate of 3.7 terawatts. Of this, 2.5 terawatts are lost through bottom friction in shelf seas. A further 1 terawatt is lost through the creation of internal tides. That leaves 0.2 terawatts which we can reasonably suppose are lost through frictional effects in the tidal flexing of the solid Earth.

The Earth's heat engine

Because our planet is a sphere, the Sun's rays fall more obliquely on the surface at high latitudes than they do at the equator. There is more solar heating per unit area at low latitudes, and the atmosphere and ocean are working constantly to re-distribute this heat. Without these fluid heat pumps, the temperature gradient from the equator to the poles would be more extreme and the Earth's climate more hostile to life.

Present estimates have it that the atmosphere and ocean work about equally hard in re-distributing heat to high latitudes, but the comparison is a difficult one to be sure of. There is no doubt, however, that ocean currents make a vital contribution to moderating global temperatures. The Sun warms the global ocean at the equator, and this warm water flows towards the poles, driven largely by the wind. For dynamical reasons to do with the conservation of angular momentum (as first reasoned by the US oceanographer Henry Stommel in the 1940s), these poleward flows hug the western boundaries of the oceans (in both hemispheres).

The western boundary currents have familiar names, for example the Gulf Stream in the North Atlantic and the Agulhas in the southern Indian Ocean. As the currents flow towards colder latitudes, surface winds carry heat from the ocean to the land, providing warmth for the continents and gradually cooling the sea. The cooled water that reaches high latitude then returns to the equator by two routes. One is as weak surface currents down the eastern boundaries of the oceans. The other is as a deep bottom flow forming part of the thermohaline circulation of the ocean (so-called because it is driven by differences in temperature and salinity).

We have sketched a simple picture of the thermohaline circulation in Figure 31. The Sun warms a relatively thin surface layer; heat is carried polewards in this layer by the wind-driven western boundary currents. As these currents approach polar regions, the water has lost enough heat to be very cold and very dense. If—in addition—sea ice forms, salt is excluded from the ice, increasing the salinity of the remaining water and further increasing its density. In a few places surface water at high latitudes in winter becomes dense enough to sink to great depths in the ocean. This happens primarily on the continental shelves of Antarctica and in the Greenland and Norwegian Seas.

Oceanographers measure the size of ocean currents in units called sverdrups (or Sv). The sverdrup, named after the Norwegian oceanographer Harald Sverdrup, is a flow of one million cubic metres of water per second. To place this in context, the flow in all the rivers in the world adds to about one sverdrup. The rate at which water is added to the deep ocean in the high latitude cooling zones is thought to be in the range 25–30 sverdrups.

The sinking water mixes a little with surrounding water as it falls, but reaches the ocean floor with temperature close to 0

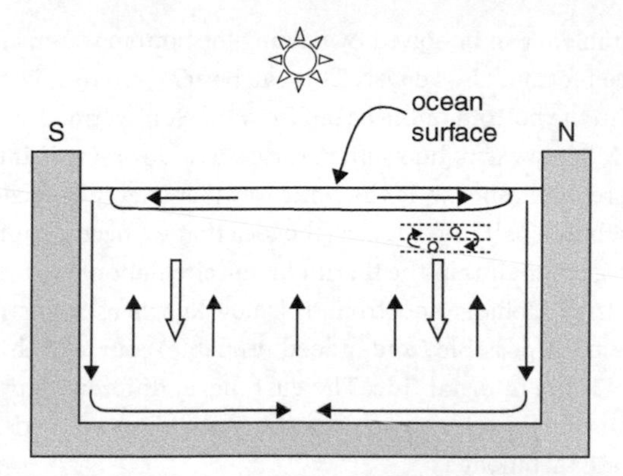

31. Sketch of the ocean thermohaline circulation. The Sun's heat, put into the surface of the ocean at low latitudes, travels polewards in surface, largely wind-driven, currents. Water sinks at high latitudes and fills the bottom of the ocean, gradually upwelling back to the surface (thin vertical arrows). To maintain this circulation, it is necessary for heat to diffuse downwards (fat vertical arrows), a process thought to be driven by vertical turbulent exchange associated with internal tides.

degrees Celsius. It then spreads throughout the ocean basins filling the bottom of the ocean, as far as the equator, with cold water. The water filling the deep ocean returns to the surface as an upwelling current throughout the world's ocean. The speed of the upwelling flow can be estimated by dividing the rate at which bottom water is created (25–30 sverdrups) by the surface area of the global ocean. This works out to be about 1 centimetre per day.

To keep the thermohaline circulation going, the cold water formed at the surface at high latitudes must be dense enough to sink to the ocean floor. This presents a problem because the bottom water has been made from the cold polar surface water: if nothing else is happening they will have the same density. The deep ocean will eventually fill with water of this density and the circulation will stop.

The problem can be solved by making the bottom water, after it has been formed, less dense. This can be achieved by mixing heat down to the bottom of the ocean from the Sun-warmed surface layer. Mixing warm, buoyant, surface water downwards into the ocean requires energy, in the same way that energy is needed to push a beach ball into the sea (the idea that we need a source of energy to maintain the thermohaline circulation was first proposed by Johan Sandström. It is now known as Sandström's theorem). A possible (and indeed probable) source of this energy is the internal tide. The currents at different depths in the internal tide slide over each other at different speeds and generate turbulence.

The turbulent eddies, moving in the vertical, transfer parcels of water between different depths in the thermocline. An upward moving parcel of water will bring cold water from the deeps and a downward moving parcel will carry warm water to a deeper layer. The exchange of water parcels between layers produces a net transfer of heat downwards. The turbulent mixing created by internal tides can provide the necessary mechanism for the vertical mixing of heat needed to maintain the ocean's thermohaline circulation. It is, indeed, difficult to see what else could provide this mechanism.

The power required to warm the deep ocean and maintain the thermohaline circulation is surprisingly small: around 10^{-3} watts for a column of water of surface area 1 square metre and stretching from the surface to the floor of the ocean (this could be provided by one kitchen food mixer whirring away in each cubic kilometre of ocean). Multiplying this power requirement by the surface area of the ocean gives the total power needed to maintain the thermohaline circulation as 0.4 terawatts. The rate at which energy is being put into the internal tide is reckoned as 1 terawatt, so it is possible for the internal tide to provide the necessary mixing energy if it runs at 40 per cent efficiency. The efficiency of conversion of mechanical energy to vertical mixing is likely not as high as this, though (it is thought to be just 20 per cent). The difference might be accounted for by allowing for some wind stirring of the deep ocean. An input of wind power of 1 terawatt, coupled with internal tide power of 1 terawatt, both acting at an efficiency of 20 per cent would provide the necessary mixing.

Chapter 7
Tidal mixing

We saw in the last chapter how internal tide waves mix the interior of the deep ocean. In this chapter, we describe some important examples of tidal mixing in shelf seas, where the water is shallow and tidal currents can be much faster than in the deep ocean.

Turbulence and mixing

Most of the energy lost from the tide through friction is first converted into turbulence: a random movement that can be seen when watching water flow. In some places the flow is fast and in others slow; there may be places where the water moves in the opposite direction to the main flow. These variations in speed are caused by eddies: swirls of water added to the mean current. The largest eddies fill the space available: the width or the depth of a channel. They break down into smaller eddies, and then smaller ones again, down to the smallest eddies in which the tide's energy is ultimately dissipated as heat. This cascade of turbulent eddies of different sizes was memorably described by the British polymath Lewis Fry Richardson:

> Big whirls have little whirls which feed on their velocity, and little whirls have lesser whirls and so on to viscosity.

One of the most obvious manifestations of marine turbulence is the great swirl of a tide-generated whirlpool or *maelstrom*. Maelstroms are often associated with narrow, tidal sea straits, between islands, for example. Differences in tidal elevation at the ends of the strait create surface slopes which drive fast 'hydraulic' currents through the strait. Where such a current emerges from the strait and interacts with other currents or topographic features, areas of lateral current shear occur and eddies, or vortices, may develop.

The whirlpool (or mythological sea monster) Charybdis terrified the ancients, as recounted by Homer in *The Odyssey*. It is sometimes linked with the modern-day Garofalo feature of the Strait of Messina, between Italy and Sicily. The Moskstraumen maelstrom off the Lofoten Islands in Norway is mentioned in old Norse poems and on nautical maps, and features in Edgar Allan Poe's 'A Descent into the Maelstrom', Jules Verne's *Twenty Thousand Leagues under the Sea*, and Herman Melville's *Moby Dick*. Some maelstroms can be very striking (Table 5), but over the years authors and artists have tended to exaggerate somewhat (see Figure 32). The largest ones are certainly not sensible places

Table 5. (In)famous tidal whirlpools or maelstroms

Whirlpool/ maelstrom	Location
Saltstraumen	Strait connecting Saltfjord and Skjerstadfjord, near Bodø, Norway
Moskstraumen	Between the islands of Moskenesøya and Mosken, Lofoten Archipelago, Norway
Corryvreckan	Between the islands of Jura and Scarba, Scotland
Naruto	Between Naruto (Tokushima) and Awaji Island, Japan
Old Sow	Between Deer Island (New Brunswick, Canada) and Moose Island (Maine, USA)

32. An illustration of a tidal whirlpool or maelstrom by Arthur Rackham (1867–1939).

to take a small vessel at peak flow, but the origins and dynamics of large whirlpools remain poorly understood.

The size of the smallest eddies—the scale at which the energy is ultimately dissipated as heat—was determined by the Russian mathematician Andrey Nikolaevich Kolmogorov using dimensional analysis. The *Kolmogorov Scale* depends on the molecular viscosity of the fluid (seawater in our case) and the rate at which the turbulent energy is dissipated. For tidal flow in shallow water it is possible to get a good idea of this second quantity. The frictional force depends on the square of the flow speed and the energy dissipation rate on the cube of the speed. For a tidal current of 1 metre per second in water 20 metres deep, for instance, the energy dissipation rate is 0.00125 watts per kilogram of water and the Kolmogorov Scale is 0.3 millimetres. If the flow increases to 2 metres per second in the same depth of water, the Kolmogorov Scale is reduced to less than 0.2 millimetres. Increasing tidal energy creates eddies which can squeeze into smaller spaces.

Turbulent eddies make a very effective mixing mechanism. One way to visualize how this happens is to imagine two adjacent volumes in the sea which, every second, exchange some of their water. Water properties, such as temperature or salinity, will also be exchanged and any differences between the two volumes will be smoothed out over time. Alternatively, we can imagine two small parcels of water initially close together being carried along by the mean flow. Because of the turbulence, they will not follow the same path exactly; the parcels will move apart as time goes on. One hundred such parcels starting close together will spread out, or diffuse, into the surrounding water.

Heat storage in the sea

An important example of mixing by the tide is the downward stirring of the Sun's heat. The sea is a great store of heat: water has a large thermal capacity and the heat stored in summer is released in winter, smoothing out the seasons. To maximize the storage, the heat needs to be mixed down. Unlike the atmosphere (which is warmed mainly at the bottom, where it touches the ground) seas are warmed at the surface. It takes energy to mix the warm, buoyant surface waters down into the colder and denser water below. In shelf seas, the principal source of this energy is the turbulence made by the wind rubbing on the sea surface and the tide rubbing on the seabed.

We can use an energy argument to work out how deep the tides can mix the Sun's heat. If water of depth *d* (from surface to bottom) is warmed and mixed so that the temperature remains uniform throughout, thermal expansion will raise the centre of gravity and the potential energy will increase at a rate proportional to the heating rate times *d*. This extra potential energy is provided by the turbulence. The rate at which tides create turbulent energy is proportional to the cube of the current speed, u^3. We can imagine that a proportion of this tidal turbulent energy is converted to potential energy as it mixes heat down. If this proportion (and the heating rate) is constant then the maximum depth of water that can be mixed by the tide, d_{max}, will be proportional to u^3. Observations confirm this idea and suggest that d_{max} in metres is about $70u^3$ if u is measured in metres per second at spring tides. This allows us to calculate the proportion of the tide's turbulent energy used for the vertical mixing of heat; it is about 0.5 per cent.

Tidal currents with maximum speed 1 metre per second are therefore capable of maintaining mixed conditions, in a mid-latitude summer, in water depths up to 70 metres. A small increase in current speed to 1.13 metres per second will mix heat down to the bottom in 100 metres of water. The most vigorously stirred shelf region that we have visited is the north channel of the Irish Sea, lying between Northern Ireland and Scotland. Here, mixed conditions are maintained all year in depths over 200 metres. Because mixing heat downwards keeps the surface cool, less heat is lost to the atmosphere. The summer heat stored in these deep mixed waters is about twice that in water without tidal stirring, a fact that must contribute to the mild winters experienced in this region.

Tides and the weather

It is sometimes possible to see, directly, the effect of tidal mixing on the temperature of the sea. Figure 33 shows the tidal curve and water temperature in winter at a coastal site in North Wales. There is a variation of temperature in phase with the spring–neap tidal cycle (the fortnightly cycle of temperature is much more pronounced than the small daily changes). The water at spring tides is nearly 2 degrees Celsius warmer than at neaps.

Observations like this are intriguing because they are hard to explain. Why should the water be warmer at spring tides? A likely explanation is that the flood tide brings, in winter, warmer offshore water to the coast, where it mixes with coastal water and raises its temperature. The effect will be greatest at spring tides because a greater volume of offshore water will be added to the mixture. If this is correct, we would expect the effect to reverse in summer when the deeper offshore waters are cooler than coastal water, and this is indeed what happens. This explanation may not be the whole story though; it is possible that local effects, involving the phase of the tide relative to the daily variation of solar heating, are also important.

Even without a full explanation of the observations in Figure 33, the fact remains that coastal water temperature can change with the spring–neap tidal cycle. If the water temperature is varying in this way, then air temperature (and possibly other meteorological parameters, such as the wind) will tend to follow suit. Experienced seafarers that we have spoken to are convinced that there are links between the tides and the weather.

Tidal mixing fronts

In parts of the shelf where the tides are not strong enough to mix the Sun's heat right down to the seabed, the water in summer becomes stratified. A surface layer warmed by the Sun and stirred

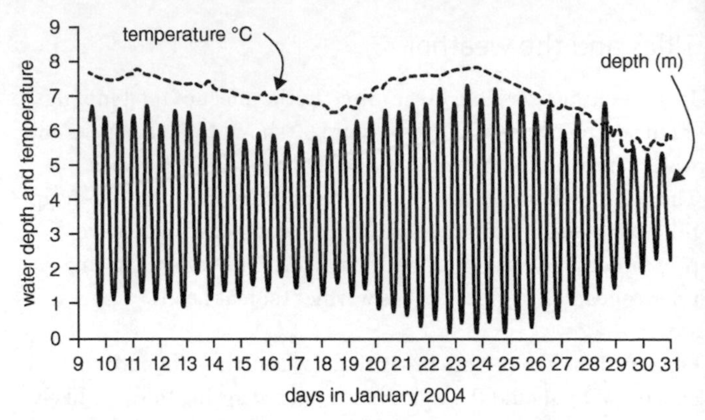

33. Variation of water temperature and depth in the Menai Strait, North Wales.

by the wind lies over a deeper, cold layer, the two separated by a vertical gradient of temperature called the seasonal thermocline.

Shelf seas at temperate latitudes in summer are divided into regions which are stratified and those which are vertically mixed, depending on the strength of the tidal streams and the depth of water. The transition from one to the other happens rapidly and creates a structure called a tidal mixing front (see Figure 34). On one side of the front, the water is stratified; on the other it is vertically mixed with an intermediate 'cool' temperature. The depth of water at the front is the maximum depth that can be mixed by the tide, and so, in metres, is about seventy times the cube of the spring current (in metres per second) at the front.

Tidal mixing fronts have been found on most of the continental shelves known to have large tides, including the Patagonian Shelf in South America, George's Bank in Canada, Cook Strait in New Zealand, and the north-west European Shelf. Unlike weather fronts in the atmosphere, tidal mixing fronts form in the same place each year. The fronts form in early spring as the

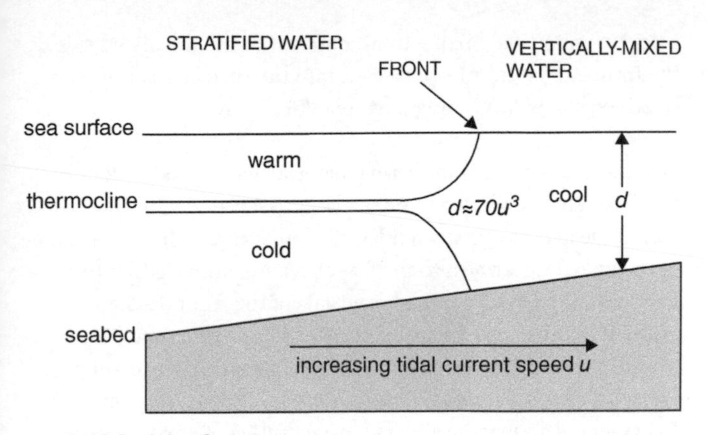

34. A tidal mixing front.

stratification forms, remain in place throughout the summer, and then gradually retreat into deeper water in the autumn as surface heating is reduced and winds pick up.

We have learned much about the behaviour of tidal mixing fronts from satellite data. Satellites equipped with infra-red sensors are able to pick out the temperature contrast between the warm surface of the stratified water and the relatively cool mixed water.

The fronts are known for their biological productivity. They create favourable conditions for the growth of phytoplankton, the microscopic algae which are the base of the marine food chain. Photosynthesis and the growth of phytoplankton are usually limited by the supply of either sunlight or nutrients (their other requirements, carbon dioxide and water, are plentiful in the sea). In the stratified parts of shelf seas, the plankton strip the nutrients out of the surface layer during their spring growth 'bloom'; the bottom layer (which has plenty of nutrients left) is too dark for them. Growth then slows down in the stratified water (it never really gets going in the mixed water because the phytoplankton, mixed from surface to bottom, receive too little light averaged over a day). Near a tidal mixing front, nutrients from the mixed side of the front can be transferred across into the surface layer of the stratified water, sustaining phytoplankton growth.

One mechanism for transferring material across a tidal front is the fortnightly variation of tidal energy associated with the spring–neap cycle. At neap tides, tidal mixing is reduced and some of the mixed water next to the front becomes stratified; as this happens nutrients from the mixed side of the front become trapped in the newly formed stratified water. Other material is mixed across as the fronts advance and retreat with a fortnightly cycle. Oxygen is depleted in the bottom layer by respiration. Instruments on marine gliders have recently detected a twice-monthly variation in oxygen concentration in the bottom layer of the stratified water close to a front in the Irish Sea, presumably created as water from the mixed side of the front (rich in oxygen because it is in contact with the atmosphere) is transferred across the front at a neap tide.

Muddying the waters

Tidal currents flowing over the seabed lift sediment particles, in large numbers, and mix them up towards the surface. The result is that tidal waters are often cloudy and opaque, no great encouragement for a dip in the sea. Observations of the particles under a microscope show that they are a mixture of flakes of mud and small living and dead organisms. Different kinds of particle stick together in clumps called aggregates or flocs, bound by biogenic polysaccharide glue. The aggregates are denser than water and tend to sink gradually towards the seabed. The Irish physicist George Gabriel Stokes showed that the sinking speed of a particle in water depends on the square of its diameter. Large flocs will settle out of suspension more quickly than individual particles.

These denser-than-water particles can be held in suspension in the sea by turbulence. The process is analogous to the downward mixing of heat: in both cases work needs to be done against gravity. We can again use an energy argument to work out how much material the tide can hold in suspension. The energy needed per unit time depends on the product of the particle concentration c and the settling speed w. If a fixed fraction of the total energy dissipated by the tide (proportional to u^3) is used to hold the particles up, then the concentration c will be proportional to u^3/w.

This idea can be tested using observations from satellites. The particles near the sea surface scatter sunlight, some of which makes it back out through the atmosphere and, remarkably, can be picked out by satellite instruments. Figure 35 shows an image of light scattered by particles in the Irish Sea. The areas of high 'brightness' (and particle concentration) marked A, B, and C are also areas of strong tidal streams—M_2 surface currents peak at more than 1 metre per second at each of these places. In fact, with the exception of some small areas of high turbidity associated with river input, the distribution of suspended particles in the Irish Sea closely matches the distribution of tidal current speeds.

In principle, the relationship between suspended particle concentration and tidal energy can be used to estimate particle settling speed, an important but difficult-to-measure parameter. Recent observations have shown that the settling speed is greater in summer than in winter. This finding is consistent with greater particle aggregation in summer when biological 'stickiness' is more active.

Isolated turbidity maxima

The patches of muddy water associated with fast tidal currents, labelled A–C in Figure 35, present something of a puzzle. There is no obvious source of fine sediments to maintain these isolated turbidity maxima, as they are called. The seabed is composed of pebbles, mostly, and there are no large rivers nearby. Without a local source, the particles in the maxima will diffuse away, down the concentration gradient into the surrounding water. A similar thing would happen if, say, a passing tanker were to discharge a cargo of ink at one of these sites. The ink would initially colour the water but then mix with its surroundings, become diluted, and eventually disappear. The satellite pictures, however, tell us that the turbidity maxima are always present despite the absence of an obvious source.

The answer to this puzzle relies on the fact that flocculated particles can change their *size* with the tide. This happens because the flocs are too frail to withstand the disrupting effects of turbulence. The upper limit for the size of a marine floc is the size of the smallest turbulent eddies—the Kolmogorov Scale. If the tidal currents increase and the turbulent scale decreases, the size of the flocs is reduced.

Within the strong currents of a turbidity maximum the flocs are small and easily held up near the sea surface where they scatter sunlight. These small flocs diffuse out of the maximum into the less turbulent surrounding water where the Kolmogorov Scale is greater. Here, the flocs grow in size (this happens by particles colliding and sticking together). The larger flocs created around the edge of the maximum now diffuse down the concentration gradient of large flocs, that is *back into the maximum*, where they are torn up into small flocs and the cycle continues.

This interpretation of the processes happening in isolated turbidity maxima has been confirmed by observations on the edge of the maximum marked A in Figure 35. These have shown that, averaged over a tide, there is a flux of large particles entering the maximum and a balancing flux of smaller ones leaving.

Tidal mixing in estuaries

Estuaries are places where salt and fresh water mix (this is true in most cases. There are estuaries in hot, dry countries where there is

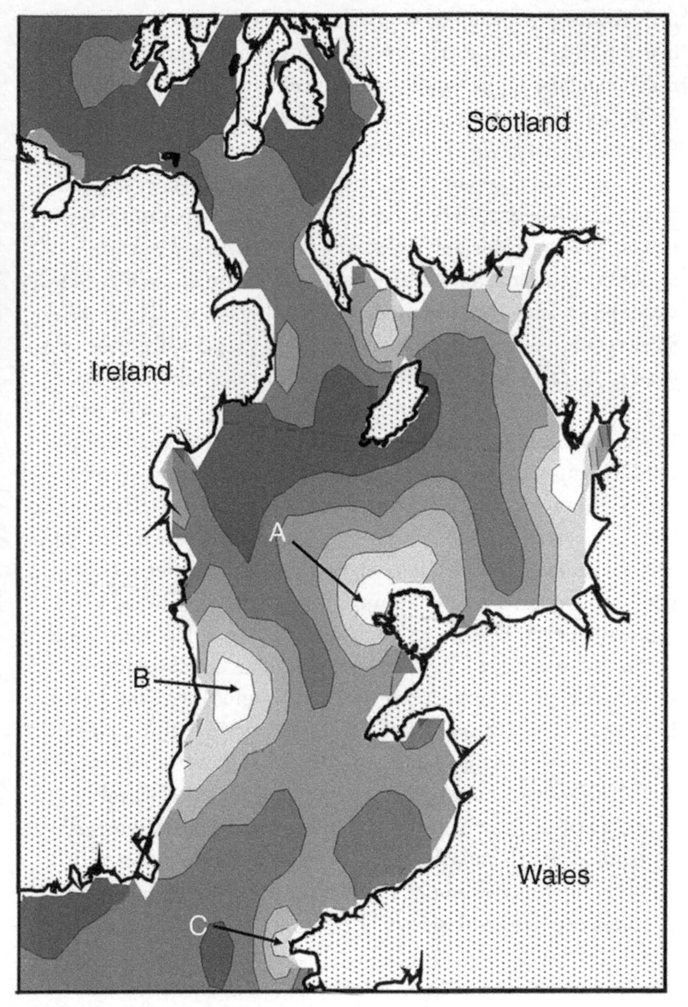

35. Brightness of the Irish Sea measured from space. Lighter shades correspond to brighter waters.

little fresh water available. In these *inverse* estuaries, the water becomes hyper-saline and processes are somewhat different). In regular estuaries with small tides, the circulation is controlled mostly by the fact that salt water is a few per cent denser than fresh water. River water entering the estuary flows over the top of salt water that has come in from the sea. A two-layer system is formed, with an interface, called the halocline, separating the layers. As river water flows seaward at the surface it pushes against the lower layer, which takes on the form of a wedge of salt water protruding into the estuary along the bottom. These *salt wedge* estuaries are found in places where the tides are small; an example is the Mississippi River in the United States.

In shallow estuaries with moderate tides, turbulence generated by tidal streams (and winds) mixes water vertically between the layers and the estuary is classified as *partly mixed*. The halocline becomes blurred and the difference in salinity between the layers is reduced. Because the outward-flowing surface layer is now partly salt water, a greater volume of water flows seaward to remove the river water. For example, if a volume V of river water enters the estuary in a day and mixes with an equal volume of seawater, then a volume $2V$ of the mixture must leave the estuary. In this case, mixing doubles the surface flow. To maintain the volume of water in the estuary there is also a flow (of volume V per day) of salt water inland along the bottom. This two-way flow is a characteristic of partly mixed estuaries and is called the *estuarine circulation*. Partly mixed estuaries are common in places with medium tides. For example, many large estuaries in northern Europe are partly mixed.

In an estuary with very large tides, the layering is destroyed completely and the estuary becomes well-mixed in the vertical with no discernible difference in salinity from surface to bottom. The estuarine circulation continues but now the rapid exchange of parcels of water between the surface and bottom slows down the horizontal flow in both directions. The effect is similar to that in the entrance to a theatre when people are leaving after a show and the next audience is coming in. As long as things are organized with people leaving the door on one side and entering on the other, the exchange can proceed smoothly. But if some people now start stepping sideways across the doorway, taking their momentum with them, the motion becomes confused and the exchange slows down. In a *well-mixed* estuary, the vertical exchange of horizontal momentum has the effect of increasing the stickiness or viscosity of the water and slowing down the horizontal flows.

The effect of bottom and side friction on the tide creates other effects in estuaries. On the flood tide, when water flows into the estuary, the faster currents at the surface carry salt water from the sea over the top of fresher water in the estuary. Because salt water is denser, this is an unstable situation. The salt water sinks, creating convection currents which add to the turbulence. In contrast, on the ebb, the faster surface currents push fresh water over salt water. This is gravitationally stable and the water can become stratified at this time. The temporary stratification created by the action of vertically sheared currents on a horizontal salinity gradient is known as *tidal straining*.

In well-mixed estuaries during the flood tide, a foam line can sometimes be seen running along the middle of the estuary (Figure 36(a)). This foam line is caused by the flood tide being slowed down by friction with the sides of the estuary. The flood current flows fastest along the central axis of the estuary. In a cross-section of the estuary during the flood, the salinity is greatest at a point in the middle at the surface. The denser salt water at this point sinks and is replaced by water flowing in from the sides (Figure 36(b)). The currents converging on the centre line carry floating material which accumulates to make the foam line, or axial convergence.

36. (a) A foam line down the middle of a well-mixed estuary on the flood tide; and (b) a sketch of the process creating the axial convergence.

The presence of the *secondary circulation* at right angles to the main flow can be demonstrated by placing a line of floats across the estuary (F to F' in Figure 36). The floats are carried upstream by the flood tide but at the same time converge on the centre line. This simple measurement enables the speed of the secondary circulation to be determined. It is surprisingly fast—up to one quarter of the speed of the main flood current. When the current turns to ebb, the secondary circulation stops and the axial convergence disappears.

Harnessing the tide

People living by the sea quickly learned to take advantage of the special opportunities that tides provide. An early example was the construction of fish traps: walled enclosures which filled with seawater (and fish) as the tide came in. In the Middle Ages in Europe, tide mills used the power of tidal streams to make flour for bread. Some of these mills still exist and are interesting places to visit.

Today, the emphasis is on using the tide to generate electricity. A number of tidal power plants are in operation worldwide and the use of the tide in this way is likely to grow. Countries with large tides have the potential to generate a significant proportion of their energy needs if the tide can be harnessed effectively.

Tides can be used to make electricity in a number of ways. One is to make a dam across the mouth of an estuary or bay. The tide flows in and out of the estuary through a turbine in the dam; the turbine is connected to a generator and electricity is produced. The power output of such a scheme depends on the square of the tidal range (since the potential energy released depends on the mass of water and the height of the centre of gravity, both of which are proportional to the range of the tide). Places with a large tidal range, such as the Severn estuary in the UK, are preferred sites for tidal dams. There is an environmental cost to schemes like this, however. The dam raises the mean water level inside the estuary and mudflats, important feeding grounds for birds, become permanently submerged.

Alternatively, electricity can be made by placing a turbine and generator directly into a fast flowing tidal stream. Some clever ideas have been proposed for maximizing the energy that can be extracted. One is a tethered glider which soars backwards and forwards at right angles to the flow (like a kite in the wind), increasing the speed at which the water flows through the turbine. Tidal stream turbines in a semi-diurnal tide will have peak generation four times a day, at times of maximum flood and ebb. They can be placed in positions around the coast such that the different phases of the tide produce electricity continuously over twenty-four hours.

Innovations in using tidal energy are encouraged by local needs and initiative. In Mozambique, students at the School of Coastal Marine Science are required to design, build, and operate a turbine to generate a small amount of electricity. The simplest rotor design has a vertical axis between curved blades making an 'S' shape viewed from above. This design, a Savonius rotor, always presents one concave and one convex blade to the flow, and since the drag is greater on the concave blade, the rotor turns at a speed which depends on the difference in the drag on the blades.

When a dynamo is attached to the turbine, something interesting happens. The dynamo can be driven by a loop of rope passed around the rotor. Power generation is equal to the product of the tension in the rope W (called the 'load', and set by the specification of the dynamo used) and the speed v with which the rope moves. However, the load and the rotor speed are inversely related to each other. A light load allows the rotor to turn quickly, but generates little power because W is small. Too much load also generates little power because the rotor turns slowly and v is small. There is an optimum load which produces the maximum value of Wv and generates the most electricity (Figure 37).

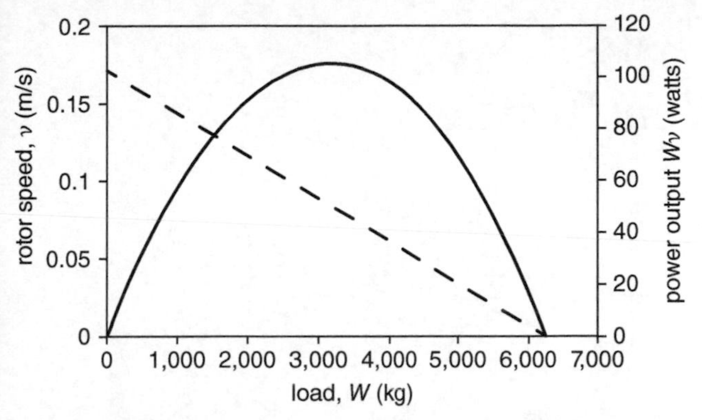

37. Power output (continuous line) and rotor speed (dashed) for a small tidal electricity plant.

For the turbine in Mozambique, which is 2 metres across and 2 metres high placed in a tidal river with maximum currents of 1 metre per second, the optimal power output is 100 watts every six hours. It's not much, but it is enough to charge batteries for a back-up power supply.

Chapter 8
New frontiers

Where lies the future for tidal studies? On our own planet there are discoveries to be made in difficult-to-reach places such as deep-sea ecosystems. The interaction between tides and sunlight in shallow water has barely been explored. Innovative computer models allow us to reproduce the tide in the early ocean. Tidal forces are not confined to Earth. Tidal flexing of the icy moons of Jupiter appears to have created a liquid water ocean on Europa. It is possible that this ocean has the right conditions for life and space probes planned for the next few years will be able to test this possibility.

Tides and deep-sea ecosystems

The effects of the tide on marine plants and animals near the coast are relatively well understood: the shallow seas of the continental shelf are easily accessible for scientific research. The same cannot be said of the deep sea. We arguably know more about the surface of the Moon, Mars, or Venus; we have certainly mapped these environments in finer detail. The deep sea is vast. It is also challenging and expensive to sample, and our understanding of how tides shape deep-sea ecosystems is accordingly poor.

In the past few decades, cold-water coral reefs (Figure 38(a)) have grabbed the attention of the scientific community. These

38. (a) A cold-water coral reef; and (b) an Arctic sponge ground.

deep-water reefs are built by 'non-zooxanthellate' corals, which do not have the symbiotic algae common to their shallow, tropical counterparts—it's too dark for photosynthesis at the depths at which they live. The classic example is the scleractinian coral *Lophelia pertusa*. Widely reported across the North Atlantic ocean, *L. pertusa* forms impressive 'carbonate mounds' to the west of Ireland and Scotland with heights up to 380 metres above the seabed and widths of several kilometres. Off Norway, it forms dense reefs on the continental shelf and in fjords. These mounds and reefs create hotspots of abundance and biodiversity in waters that might otherwise seem devoid of structural habitat and life.

Cold-water coral reefs are now fairly well studied and one thing is clear: tides are important. Corals are passive 'suspension feeders', extracting food particles from the water as it flows by. Tidal currents are often accelerated where they interact with prominent features on the seabed and this enhances the supply of food (and larval recruits) to the reefs. It's easy to imagine how currents amplified from typical deep-sea speeds of just a few centimetres per second to speeds of several tens of centimetres per second can benefit such organisms. Even hungry corals can have too much of a good thing, of course, and tides help here too, ensuring they don't get smothered by heavy loads of suspended particles settling from above.

Internal tidal motions (Chapter 6) are also thought to play an important role. They enhance turbulence in the water column, mixing nutrient-rich deep water upwards, and potentially enhancing productivity over (and food supply down to) some reefs. They intensify near-bed currents, whipping up particulate material (potential coral food) from the seabed into layers of enhanced turbidity or 'nepheloid layers' (from the Greek *nephos*, meaning cloud) and carrying them past the corals. Tidally driven internal hydraulic jumps can occur over coral reefs, causing regular, rapid injections of plankton-rich surface water downwards (known as 'downwelling'). Tidal currents are even believed to control the precise height and shape of the carbonate mounds.

More recently, the spotlight has fallen on *deep-sea sponge grounds* (Figure 38(b)). The sponges are a very diverse, mostly marine group of animals with a relatively simple lifestyle that has proved successful since the Cambrian Period (541–485 million years ago). As adults, they live attached to the seabed (i.e. they are 'sessile'). Most possess chambers of flagellated cells, the undulatory beating of which induces currents in an internal system of channels from which they extract their food (i.e. they 'filter feed', often processing impressive volumes of seawater). Some sponges also have their 'exhausts', or *oscula*, elevated (like chimneys) into the faster-flowing water away from the seabed, which serves to lower the pressure at the oscula relative to that at the water intake openings, or *ostia*, positioned lower down on the sponge. This body design creates a pressure gradient that passively drives additional flow through the sponge (from high to low pressure, or *ostia* to *oscula*), supplementing the active water pumping and conserving precious metabolic energy.

Several sponge species aggregate into grounds thought to present similar ecological benefits to cold-water coral reefs. The 'Bird's Nest' sponge (*Pheronema carpenteri*) forms grounds along the European and north-west African continental shelves and slopes, and off the Azores archipelago. Aggregations of the 'Russian Hat' sponge (*Vazella pourtalesi*) are found off Canada's east coast. Nine-thousand-year-old sponge reefs have been discovered off Canada's west coast, and extensive sponge grounds occur under the ice shelves of Antarctica. Like the corals, many sponges undoubtedly benefit from some level of intensified flow and enhanced food supply. Debate still rages over the exact conditions they prefer: sponges, after all, can become 'clogged up'.

At the Arctic Mid-Ocean Ridge, some seamounts host sponge grounds dominated by massive demosponges (resembling giant puffball mushrooms) and delicate, trumpet-shaped glass sponges. On one recently sampled seamount, the density and diversity of sponges reached a clear and impressive peak at the depth of a boundary between water masses, which was very dynamic owing to internal tidal motions. Rather frustratingly, we could only speculate on the factors most important to the sponges, based partly on what we know from cold-water coral and other deep-sea research. We may understand a lot about how the tides structure and enhance familiar ecosystems at the air–sea interface, but there is much to learn about how internal tides and tide–seabed interactions affect the enigmatic ecosystems of the deep.

Tidal interactions with sunlight

Almost all marine ecology textbooks refer to the importance of the tide in numerous shallow-water settings and contexts. It influences the horizontal banding of organisms, the *vertical zonation*, we see on the seashore and dominates the rhythms of everyday life, the *chronobiology*, of the intertidal and subtidal zones (see Chapter 1). You are unlikely, however, to read about the tide's interactions with sunlight, despite several important ecological implications.

Sunlight entering the sea is attenuated with depth (Figure 39). The extent of the attenuation depends on the concentrations of light-absorbing and scattering materials present. If a constant fraction of the light is 'lost' in each metre, light energy will decay exponentially. The phenomenon is described mathematically by what's known as the Lambert–Beer Law and the rate of attenuation with depth is normally represented by an 'attenuation coefficient', k_d. In a tidal sea both the water depth and k_d fluctuate regularly over time, and it is easy to see how light reaching the seabed must be affected by the tide.

For shallow, turbid inshore areas with a large tidal range, the patterns in light reaching the seabed can be very different to those experienced on land. For example, if low water coincides with midday there will be a period of high light levels at the seabed around this time. If, however, high water occurs at midday the seabed will be relatively dark instead. In this case, the low waters that precede and follow (at around 6 am and 6 pm for semi-diurnal tides) produce two distinct periods of light at the bed, providing there are sufficient hours of daylight.

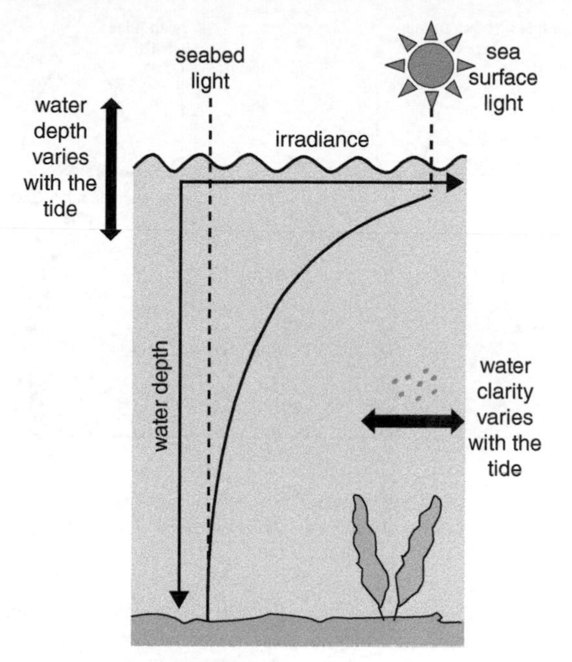

39. Tidal processes affecting seabed light.

As we noted in Chapter 1, a given location will tend to have high waters (and low waters) that always occur at the same times of day during spring tides. These times advance by about fifty minutes each day until seven to eight days later, at neap tides, the opposite scenario occurs (low waters occur at the times high waters occurred during spring tides and vice versa). A further seven to eight days sees everything back at its original time for the following spring tide. This leads to spring–neap cycles in the daily total light received by the seabed (Figure 40). It will receive more light, summed over the day, at spring tides if, for example, low water occurs at midday (and, conversely, high water of neap tides occurs at midday). The Bay of Brest in Brittany (France) is one such location. The Menai Strait in Wales (UK) has a directly opposite cycle (data shown in Figure 40); low water

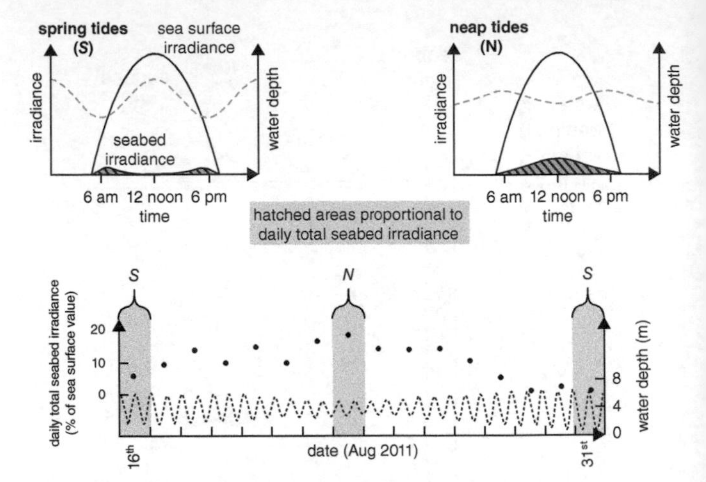

40. Daily and spring-neap patterns in seabed light at a location where high waters of spring tides occur at around midday and midnight.

occurs at midday during neap tides and the bed typically receives most light then.

At middle to high latitude sites, spring–neap cycles in daily total seabed light are affected by seasonal changes in the number of daylight hours per day. A long summertime daylength will tend to smooth out the difference in daily totals between springs and neaps: the benefits of low water occurring at midday are offset by morning and evening high waters and, conversely, the disadvantages of high water at midday are offset by morning and evening low waters. A very short winter daylength will tend to exaggerate the difference: the focus becomes much more about the water depth at around midday, which may be very different from springs to neaps.

The exponential decay of light with depth also has an important consequence: from mid-tide to low water there may be a disproportionately large gain in light at the bed (Figure 41). This

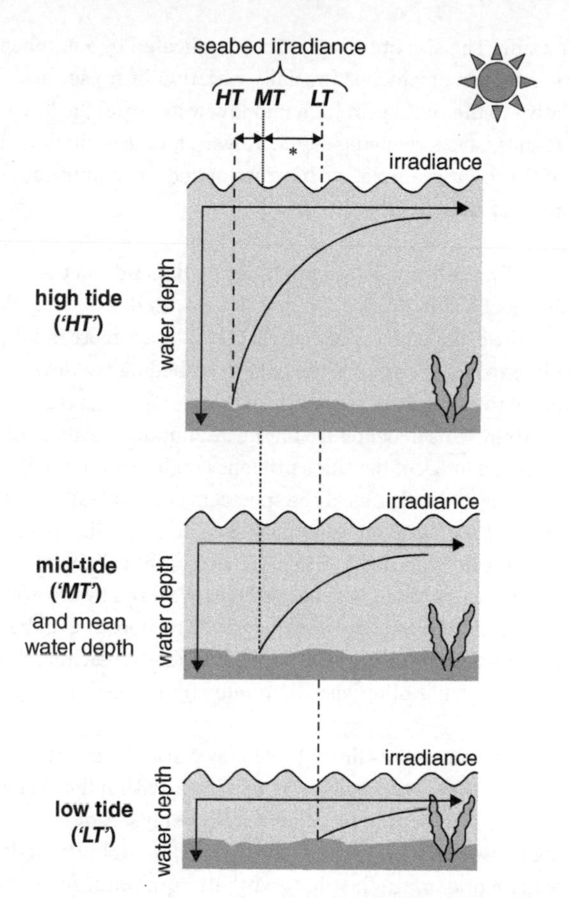

41. Tidal amplification of seabed light. Note the effect of the bed intersecting the irradiance curve at different depths (during different stages of the tide) on the light it receives.

gain is often greater than the loss of light around high water, so these periods do not necessarily 'average out' over a tidal cycle, or several, as you might expect. The result is that the tide can amplify seabed light over time, relative to a situation that is comparable but has no tide. The size of this effect is complicated by a number of factors (not least of which is the daily variation in sea surface light, relative to the timings of high and low water) and the tide can sometimes have a reducing effect. However, on certain days in the Bay of Brest, for example, we have estimated the amplification to be a factor of almost 32!

Mathematically, the amplification that occurs depends on the attenuation coefficient (amongst other factors), as this dictates the steepness of the exponential decay curve. Light decays more rapidly with depth in turbid waters and the gain in seabed light at low water relative to mid-tide is greater in such waters than it is in those less turbid. This dependence on the attenuation coefficient leads to one final quirk of the tide's influence on underwater light. Light at the blue and red ends of the spectrum (i.e. 400–500 nanometres and 650–700 nanometres, respectively) is the most strongly attenuated with depth in coastal waters (blue light by dissolved organic matter and suspended particles; red by the water itself) and is thus amplified most by the tide. The boosting of these wavelengths flattens somewhat the spectrum of light reaching the bed over time, which is otherwise dominated by green-yellow light.

It's curious to think a semi-diurnal tide may cause a piece of seaweed to effectively experience two 'daytimes' within the single day–night cycle experienced by us terrestrial beings. A more interesting consequence, we feel, is that, providing light levels do not overwhelm or damage the photosynthetic apparatus of seabed plants and algae, and they are not limited by other factors, spring–neap patterns in daily total seabed light may drive fortnightly cycles in their photosynthesis and growth. Imagine shallow-water marine ecosystems, such as kelp forests and seagrass meadows, pulsing with enhanced primary productivity every two weeks.

What's more, the amplification of seabed light over time by the tide probably makes seaweeds more productive (and allows them to grow deeper) in some tidal areas than in non-tidal areas with otherwise similar conditions. The likely flattening of the time-averaged spectrum of light at the bed by the tide may also prove important to seabed algae possessing greater concentrations of photosynthetic pigments primarily capturing the boosted (blue and red) wavelengths. Perhaps in this way the tide is influencing the particular species that thrive at given depths well *below* the low tide mark, in addition to its better-known effects on those above it. Many of these ideas need further testing, but we can be sure that if we ignore the interaction of the tide and sunlight we will make poor predictions of the energetics and productivity of several marine ecosystems that may be playing very important roles in regulating our climate and sustaining marine biodiversity.

Tides in the early ocean

The study of the tide in earlier oceans is important because of, among other things, what it can tell us about the origin of the Moon. As we have seen, tidal friction is gradually pushing the Earth and Moon apart. If we can work out the rate at which the Moon has receded in the past, we can form an estimate of the age of the Moon. This delicate task requires knowledge of tides and tidal friction in times past. There is little observational evidence to help here; this is the realm of hard-core computer modellers. Moreover, the work requires a special kind of hybrid model which can allow for the effects of continental drift on the shapes and depths of the oceans over periods of thousands of millions of years.

As tidal friction slows the Earth's spin, the total energy and angular momentum of the Earth–Moon system are conserved. The energy lost from the Earth's spin is passed to the ocean as heat and mixing energy (the latter ultimately raises the potential energy of the ocean). The angular momentum is given to the Moon, and, as a result, the Moon is receding from the Earth. The length of the day and that of the month are both increasing.

The length of the day and month can be calculated from the radius of the Moon's orbit, using Kepler's third law and the principle of conservation of angular momentum. These calculations, the results of which are shown in Figure 42, don't require (or include) any assumptions about tidal friction. Starting at a time when the Earth and Moon were close together, the Earth was spinning rapidly and the daylength and the length of the month were both short. At this time, tidal friction caused the days to lengthen more slowly than the month and the number of days per month increased up to a maximum of thirty-one. The daylength then began to increase more rapidly than the month and the number of days per month began to fall. At some time far into the future, the length of the day will increase rapidly, and the day and the month will become equal. The Earth will then keep the same face towards the Moon and both bodies will be *tidally locked* to each other.

Although we can be sure of the sequence of events shown in Figure 42 it is much more difficult to put a timescale on them.

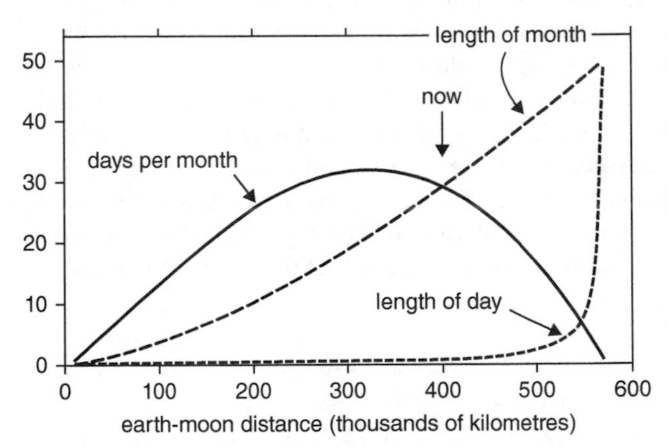

42. The length of the month and of the day (both expressed as multiples of the present day) plotted against the distance between the Earth and the Moon. The third curve shows the number of days per month, which went through a maximum sometime in the past. We are currently at the time marked by the 'now' arrow.

The English physicist George Darwin, son of the famous naturalist Charles Darwin, likened this problem to travelling on a familiar train journey having neglected to take a timetable. The stations on the journey pass in a predictable way but the exact time the train arrives at each station cannot be predicted exactly in advance.

A timetable can be made if the speed of the train is known, or in the case of the Earth and the Moon, if the rate of loss of energy from the Earth's spin can be calculated at all times in the past and the future. We now know (through analysis of Moon rocks collected by Apollo 14) that the age of the Moon is 4,500 million years: about the same age as the solar system. Is this figure consistent with the age of the Moon's orbit, starting from a time when the Earth and Moon were close together and moving forwards to the present day?

We can answer that question if tidal dissipation rates (and so the rate of Earth–Moon separation) are known for all times in the past 4,500 million years. That's a lot to ask for and at present we don't have that information. If we assume that tidal energy dissipation has always been the same as it is today, we calculate an age of the Moon's orbit as 1,500 million years—a factor of 3 too short. It must be, therefore, that in the early ocean, tidal dissipation rates were less, on average, than they are now. There is some observational evidence to support this conclusion. Currently the Earth–Moon distance is increasing at a rate of 38 millimetres per year and the length of the day by 0.024 milliseconds per year. Sedimentary rocks laid down 620 million years ago can be used to fix the average value of these figures over that period. These are estimated to be 22 millimetres and 0.012 milliseconds per year, about half the present values.

We can seek to confirm the fact that tidal dissipation has been less in the past by running computer models of the global ocean tide. These models combine scientific disciplines. The map of the Earth was different hundreds of millions of years ago:

continents have moved over periods extending back that far and, as a result, the shape and depth of the world ocean has changed. In recent years it has been possible to combine models of continental drift with global tidal models to estimate energy dissipation over the last 250 million years. These model runs indicate that we are currently near a peak in tidal energy dissipation. The present day ocean is close to resonance with the semi-diurnal tidal forcing and the early ocean was further from resonance. Average dissipation rates over the last 250 million years were about half of what they are now. This means that the Moon receded more slowly in the past than now and allows us to increase the estimate of the age of the Moon's orbit.

Although dissipation rates were less, on average, in the past, there were short periods when the trend was reversed and there were large tides. If we move back in time to the last ice age, for example (about 12,000 years ago), the shape of the continents and the tidal forcing were much the same as they are today but there was much more ice and, as a result, sea level was lower. Crucially, the shelf seas—currently the major source of tidal friction—were dry. Tides were confined to the ocean and their energy could not be sapped by sending tidal waves into the shallow, frictional shelf seas.

We might expect, then, that tides would have been larger during the last ice age and computer simulations agree with that assessment. The tides in the Atlantic, in particular, were larger than they are today. The models also tell us that tidal friction was also greater during the last ice age than it is now. The slightly shallower oceans at that time were closer to resonance with the semi-diurnal tidal forcing. During the last ice age, despite the lack of frictional shelf seas, more energy, rather than less, was dissipated than at the current time.

As we go further back in time it is more difficult to be sure how the tides behaved but there is one particular period which has captured the imagination. The Devonian (420–360 million years ago) was a critical time in the evolution of life on Earth. The fossil record tells us that this was when animals with backbones left the ocean to live on the land. It has been speculated that the tides were crucial to this process. Fish stranded on the beach by the receding tide would have needed to develop lungs and limbs to allow them to survive until they were rescued by the tide coming back in. Gradually they would have spent longer on land and eventually left the sea behind completely. We would not be here if it were not for the tide. As we search for advanced life in our galaxy, we should therefore look for a planet with a watery ocean *and* a tide-making Moon.

Tides in the solar system

The theory so painstakingly developed for the tides of our ocean can be applied to the moons and planets of the solar system; it is also possible to make close observations of these systems to test the theory. A number of unmanned space probes have flown close to the moons of Saturn and Jupiter and more are planned (Table 6). It seems very likely that important discoveries will be made about these worlds within the next few decades.

Table 6. Exploring the moons of Jupiter and Saturn

Spacecraft	Dates	Notes
Voyager 1 and 2	1979	Flew by Jupiter and Saturn
Galileo	1995–2003	Flew by Jupiter's moons
Cassini	2004–15	Flew by Saturn's moons
JUICE	2022 (launch)	Mission to study Jupiter's three icy moons—Callisto, Europa, and Ganymede
Europa Clipper	2020s	Planned to fly by Europa multiple times

The tidal force exerted by a planet on its moon is proportional to the mass of the planet, the diameter of the moon, and the inverse cube of the distance between their centres. We would therefore expect the greatest tidal effects on large moons orbiting close to a major planet. The winner here, by some distance, is Io, the closest to its parent of the four Galilean moons of Jupiter. Io is about the same size as our own Moon and orbits Jupiter at about the same distance as that of the Moon about the Earth. But, because of the large mass of Jupiter, the tidal forces exerted on Io are over 5,000 times larger than the Moon's tidal force on Earth.

As it orbits its parent planet, Io is squeezed and stretched by the tidal force of Jupiter and that of the other Galilean moons. These tides of Io are the equivalent of the tides in the solid body of the Earth we mentioned earlier. On Io, the solid body tides are spectacular. If the material of which Io is made has about the same elasticity as that of the Earth, the tidal bulges produced will be of order 100 metres high. We can expect some dramatic effects.

Io does not disappoint. The tidal flexing of the moon causes internal heating which melts and boils the core (unlike Earth, where the interior is heated by nuclear fission, tidal heating is the principal source of heating on Io). The boiling core erupts through volcanoes on the surface sending plumes 300 kilometres into space.

Although Io experiences the strongest tidal forces in the solar system, the headlines have been grabbed by the next of Jupiter's moons, Europa. Europa has a thick coating of water ice and calculations suggest that tidal heating may be sufficient to create a liquid water ocean beneath a surface shell of ice. Observations support this idea. First, the Hubble Space Telescope and, later, the spacecraft Galileo detected what appear to be plumes of water vapour emitted through cracks in the ice cover. The cracks themselves are caused by tidal flexing, and open and close with a tidal rhythm. Water from a Europan ocean flows up a crack as it opens, sometimes bursting out of the surface as a plume of water vapour.

The cracks allow sunlight to get into the cold, but liquid, water on Europa, creating the necessary conditions for photosynthetic life. The coldness of the water is not a problem: phytoplankton on Earth thrive in polar seas. But is there enough light? Marine algae in Earth's ocean can manage on light levels about one hundred times less than those received at the surface of our planet. Jupiter is 5.2 times further from the Sun than Earth is. Sunlight decreases with the inverse square of distance from our Sun and light levels reduce from Earth to Europa by a factor of 27. Some light will be lost in the thin ice at the surface of the cracks, but there should be enough. It is also likely that any Europan algae would have adapted to survive at very low light levels.

Europa currently offers one of the best locations for our nearest living neighbours. Future space missions are planned to collect and analyse samples for evidence of life. The best way to look for this, without landing (and possibly contaminating the moon), may be to sample the water vapour in one of the plumes.

The tides of Jupiter's satellites are different to those of Earth. Both Europa and Io keep the same face towards Jupiter; the strong tidal friction has long since sapped the spin energy of the moons and they now rotate about their axes with the same period that they orbit Jupiter. The moons are said to be tidally locked to their planet (the same thing has happened to our own Moon). As a result, the tidal bulges do not travel across the surface of the moons (as they do on the spinning Earth); instead tides are created by the fact that the orbits are slightly elliptical.

The change in the strength of the tidal forcing as the distance from Jupiter varies during an orbit causes the bulges to grow and shrink (by about 30 metres in the case of Europa). It is this movement that creates the flexing that produces heat. Tidal friction works to remove the eccentricity of the moons' orbits but it is maintained in this case by the regular gravitational pull of the other large moons of Jupiter (an effect known as orbital resonance). Enceladus, the second moon of Saturn, is also thought to have a liquid ocean below an ice cover, although in this case the ice is probably too thick to allow life in the ocean.

The principles which create tides in our oceans and seas operate throughout the universe. Tidal forces are created whenever one body orbits about another. Our solar system is the closest thing we can observe to perpetual motion: the planets, comets, and asteroids orbit the Sun and the moons orbit their planets without any apparent change. It must be, however, that—gradually—tidal friction in the oceans, atmospheres, and solid bodies of the planets is sapping the kinetic energy of the solar system. The energy is transformed to heat and radiated to space, increasing entropy. These changes are so slow, however, that their consequences are unlikely to be important during the lifetime of the Sun.

Further reading

More information about the topics in the earlier chapters of this book can be found in the classic text books on tides, some of which are easily available and others which require a bit of tracking down. The ones we recommend are:

J.D. Boon. *Secrets of the Tide* (Woodhead, 2011)
D. Cartwright. *Tides: A Scientific History* (Cambridge University Press, 1999)
G.H. Darwin. *The Tides and Kindred Phenomena in the Solar System* (Houghton, 1899)
A. Defant. *Ebb and Flow* (University of Michigan Press, 1958)
A.T. Doodson and H.D. Warburg. *Admiralty Manual of Tides* (HMSO, 1941)
G. Godin. *Analysis of Tides* (Liverpool University Press, 1972)
D. Pugh. *Tides, Surges and Mean Sea-Level* (John Wiley & Sons, 1996)
D. Pugh and P.L. Woodworth. *Sea Level Science* (Cambridge University Press, 2014)

In addition, you can consult the following for more information and entertainment on the subject of specific chapters:

Chapter 1: Watching the tide

P. Caton. *No Boat Required: Exploring Tidal Islands* (Matador, 2012)
E. Childers. *The Riddle of the Sands* (Smith, Elder and Co., 1903)
M. Marten. *Sea Change* (Kehrer Books, 2012)

E. Naylor. *Moonstruck: How Lunar Cycles Affect Life* (Oxford University Press, 2015)
A.C. Redfield. *The Tides of the Waters of New England and New York* (Woods Hole, 1980)

Chapter 2: Making tides

Oceanography Course Team. *Waves, Tides and Shallow Water Processes* (Open University, 1989)
R. Feynman.Resonance. *Lectures on Physics* (Addison-Wesley, 1963)
D.N. Thomas and D.G. Bowers. *Introducing Oceanography* (Dunedin, 2012)

Chapter 3: Measurement and prediction

J.J. Dronkers. *Tidal Computations in Rivers and Coastal Waters* (Wiley, 1964)
B.B. Parker. *Tidal Analysis and Prediction* (NOAA, 2007)
P.L. Woodworth. *Three Georges and One Richard Holden: The Liverpool Tide Table Makers* (Transactions of the Historic Society of Lancashire and Cheshire, 2002)

Chapter 4: The tide in shelf seas

K.F. Bowden. *Physical Oceanography of Coastal Waters* (Ellis Horwood, 1983)
D.G. Bowers and G.W. Lennon. *Tidal Progression in a Near-Resonant System* (Estuarine, Coastal and Shelf Science, 1990)
J.H. Simpson and J. Sharples. *Introduction to the Physical and Biological Oceanography of Shelf Seas* (Cambridge University Press, 2012)—also very relevant to Chapter 7
D. Webb. *On the Shelf Resonances of the Gulf of Carpentaria* (Ocean Science, 2012)

Chapter 5: Tidal bores

S. Bartsch-Winkler and D.K. Lynch. *Catalog of Worldwide Tidal Bore Occurrences and Characteristics* (U.S. Geological Survey Circular, 1988)
H. Chanson. *Tidal Bores, Aegir, Eagre, Mascaret, Pororoca: Theory and Observations* (World Scientific Publishing, 2011)

D.K. Lynch. Tidal Bores, *Scientific American*, 247/4 (1982), 146–56
S. Pond and G.L. Pickard. *Introductory Dynamical Oceanography* (Pergamon Press, 1991)

Chapter 6: Tides and the Earth

W. Munk and C. Wunsch. *Abyssal Recipes II: Energetics of Tidal and Wind Mixing* (Deep Sea Research, 1998)
S. Rahmsdorf. Thermohaline Ocean Circulation. *Encyclopedia of Quaternary Sciences* (Elsevier, 2006)
R.D. Ray and G.D. Egbert. Tides and Satellite Altimetry. *Satellite Altimetry over Oceans and Land Surfaces*, ed. D. Stammer and A. Cazenave (CRC Taylor and Francis, 2017)
R. Soulsby. The Bottom Boundary Layer of Shelf Seas. *Physical Oceanography of Coastal and Shelf Seas* (Elsevier, 1983)
F.R. Stephenson. *Historical Eclipses and Earth's Rotation* (Cambridge, 1997)
G.I. Taylor. *Tidal Friction in the Irish Sea* (Philosophical Transactions of the Royal Society, 1918)

Chapter 7: Tidal mixing

D. Prandle. *Estuaries: Dynamics, Mixing, Sedimentation and Morphology* (Springer Verlag, 1992)
S.A. Thorpe. *An Introduction to Ocean Turbulence* (Cambridge, 2007)

Chapter 8: New frontiers

A. Friewald and J.M. Roberts (eds). *Cold-Water Corals and Ecosystems* (Springer-Verlag, 2005)
J.A.M. Green et al. *Explicitly Modelled Deep-Time Tidal Dissipation and Its Implications for Lunar History* (Earth and Planetary Science Letters, 2017)
R. Greenberg (ed.). *Unmasking Europa: The Search for Life on Jupiter's Ocean Moon* (Copernicus, 2008)
M.M. Hogg et al. *Deep-Sea Sponge Grounds: Reservoirs of Biodiversity*, Biodiversity Series No. 32 (UNEP-WCMC, 2010)
J.T.O. Kirk. *Light and Photosynthesis in Aquatic Ecosystems* (Cambridge, 1994)

M. Maldonado, et al. Sponge Grounds as Key Marine Habitats: A Synthetic Review of Types, Structure, Functional Roles, and Conservation Concerns. *Marine Animal Forests*, ed. S. Rossi, L. Bramanti, A. Gori, and C. Orejas (Springer International Publishing, 2015)

M. White, I. Bashmachnikov, J. Aristegui, and A. Martins. Physical Processes and Seamount Productivity. *Seamounts: Ecology, Fisheries, and Conservation*, ed. T.J. Pitcher, T. Morato, P.J.B. Hart, M.R. Clark, N. Haggan, and R.S. Santos (Blackwell Publishing, 2007)

Glossary

altimeter (satellite) A satellite-mounted radar system used to determine sea surface height above a fixed level, based on the time it takes for a radar pulse to travel from the satellite to the sea surface and back. Today, heights can be determined with accuracies of a centimetre.

amphidrome/amphidromic system A point of no vertical rise and fall of the tide, created when two *tide waves*, travelling in opposite directions, cancel each other at all times. The currents in the two waves, however, add at the amphidrome.

capillary waves, see gravity and capillary waves.

Coriolis effect An apparent deflection of moving objects on a rotating planet. On the Earth, the deflection is to the right in the northern hemisphere and to the left in the southern hemisphere. The Coriolis effect makes winds blow around pressure systems and creates tidal *amphidromes*.

co-tidal chart A map showing co-tidal lines (along which high water occurs at the same time) and co-range lines (along which the tidal range is constant).

deep-sea sponge grounds Mass occurrences of large sponges in the deep sea (>200 metres water depth) whereby sponges dominate the seabed fauna. Aggregations may be monospecific (one species) or multispecific.

diurnal inequality A difference in the level of the two high waters (or two low waters) on the same day.

diurnal tide A tide in which there is just one high and low water each lunar day. Purely diurnal tides are rare but are observed along some coasts, where *resonance* amplifies the diurnal *harmonics* of the tide.

entrainment In oceanography and fluid dynamics, something incorporated into a flow may be described as having been entrained. For example, air bubbles are entrained into turbulent *tidal bore* fronts, eroded sediments are entrained into tidal flows, and nutrient rich deep waters are entrained into surface waters by the vertical mixing action of *internal tides*.

equilibrium tide The tide that would occur in an ocean which covered the Earth (with no land masses present) and which was able to respond instantaneously to changes in the *tide-generating force*.

equinoctial tides The largest *semi-diurnal tides* occur at the equinoxes, in March and September, when the Sun lies in the plane of the Earth's equator.

gravity and capillary waves Water waves which move in response to the *pressure gradient force* due to the surface slope between the crest and the trough are sometimes called gravity waves, because gravity is an essential part of making the wave. In contrast, the small ripples sometimes seen on the water surface rely on surface tension, rather than gravity, for their restoring force and are called capillary waves.

harmonics The rhythms which contribute to the tide. A pure harmonic is a regular, sinusoidal, up-and-down motion of fixed amplitude and period as illustrated in Chapter 3 of this book.

hydraulic jump An abrupt rise in the water surface that occurs at the transition between two different flow regimes. In the case of *tidal bores*, a hydraulic jump occurs as a result of the discontinuity of flow at the point where the incoming tide meets the river outflow. It travels upriver with the flooding tide.

internal tide Tidal motion imposed on layers of different density in the ocean. Internal tides often form when tidal streams in stratified water flow over submarine mountains, such as those found at mid-ocean ridges.

neap tides, see spring and neap tides

pressure gradient force A horizontal force created by differences in pressure in the sea. When the sea surface is sloping relative to the horizontal it creates a pressure gradient force acting down-slope.

progressive and standing waves Water waves which travel over the seabed are called progressive waves. The speed at which they travel in water shallow compared to their wavelength just depends on the depth of water. Two equal progressive waves travelling in opposite directions create a standing wave in which there are nodes at intervals of half a wavelength: places of no vertical tide but fast tidal streams, where the elevations in the two waves always cancel but the currents add.

refraction A change of wave direction caused by a change of wave speed. Waves travelling in shallow water have speeds determined by the water depth. The parts of the wave travelling in deeper water will travel faster than those in shallower water, with the effect that the wave crest will swing round towards alignment with the depth contours.

resonance/resonant period The response of a body when it is forced at a regular period close to its natural period of oscillation. Tides in the ocean are enhanced because the oceans are close to resonance with the period of the tidal forcing.

semi-diurnal tide A tide in which high waters occur at intervals of, on average, twelve hours and twenty-five minutes, or about half a day.

shallow-water waves, or long waves, have long wavelengths compared to the depth of water in which they travel and their speed is equal to $\sqrt{(gd)}$ where g is the acceleration due to gravity and d the water depth. In contrast, waves travelling in water which is deep compared to their wavelength travel at a speed which depends on their wavelength, with the long waves outstripping the shorter ones. *See also gravity and capillary waves.*

shelf sea A sea, generally less than 200 metres deep, covering a continental shelf.

spring and neap tides The fortnightly variation in tidal range caused, in the case of *semi-diurnal tides*, by the relative positions of the Moon and Sun with respect to the Earth. When the Sun and Moon make a straight line with the Earth their tidal forces combine and large spring tides result. When the Sun and Moon make a right angle with the Earth, smaller neap tides occur. There is a spring–neap cycle in the *diurnal tide* too, but this is controlled by the declination of the Moon and Sun.

standing waves, see progressive and standing waves

storm surge An increase in sea level caused by meteorological effects, for example an onshore wind or a low pressure system. Storm surges on top of a large *spring tide* can cause serious flooding.

tidal locking The situation in which an astronomical body orbiting another takes as long to rotate about its own axis as it does to orbit its partner. The orbiting body then constantly shows the same hemisphere to its partner ('synchronous rotation').

tide waves Water waves with period equal to the semi-diurnal or diurnal tidal period and created, directly or indirectly, by tidal forces. Tide waves are long compared to the depth of the ocean and travel at a speed which just depends on the depth of water.

thermocline Depth range with a relatively strong vertical gradient of temperature, marking the transition between an upper layer of warm water and a lower layer of cold water (in a stable state). Similar terms exist for the salinity and density fields (halocline and pycnocline, respectively).

tidal bore A *hydraulic jump* that forms on the incoming tide and travels upriver in some funnel-shaped estuaries with a large semi-diurnal tidal range. A tidal bore may be undular (a smooth, non-breaking wave) or turbulent (breaking). *See also hydraulic jump* and *whelps*.

tidal prism The volume of water that flows into an estuary or bay during the flood tide.

tidal straining The action of a vertically sheared tidal flow on a horizontal density gradient. On the ebb tide in a well-mixed estuary, for example, the faster surface currents pull less dense water over denser water further out to sea creating stratification. The stratification will generally (but not invariably) be destroyed when ebb turns to flood. The term is also used to mean flexing of the solid Earth by tidal forces.

tide-generating force The tide-generating force created by the Moon on the Earth is the difference between the gravitational attraction of the Moon at a point and the gravitational attraction of the Moon at the centre of the Earth. An equivalent definition applies to the Sun's tide-generating force and indeed any two astronomical bodies. It is the horizontal component of the tide-generating force that raises tides in the ocean.

tsunami An ocean wave of very great wavelength generated by the abrupt displacement of a large volume of water by, for example, seismic disturbances (earthquakes), submarine landslides, volcanic eruptions, comet/asteroid impacts, glacier calving, etc. Tsunami are not tidal phenomena and should not be referred to as 'tidal waves'.

turbulence A random motion which causes fluctuations in the velocity of a tidal stream or other flow and which transfers parcels of water between different parts of the flow. Turbulence derives its energy from the mean flow and the energy is ultimately dissipated as heat in the smallest turbulent eddies.

water mass A body of water with a distinctive range of temperature and salinity, which is identifiable based on these and other physical and biogeochemical properties. Water masses acquire their characteristics at the sea surface owing to climatic conditions at particular locations. They sink and spread along an appropriate density surface where they can be traced based on their properties.

whelps A train of secondary waves, typically of smaller amplitude, following the lead wave of a *tidal bore*, particularly one of undular type (*see tidal bore*).

Index

For the benefit of digital users, indexed terms that span two pages (e.g., 52–53) may, on occasion, appear on only one of those pages.